Methods in Molecular Biology™

Series Editor
John M. Walker
School of Life Sciences
University of Hertfordshire
Hatfield, Hertfordshire, AL10 9AB, UK

For further volumes:
http://www.springer.com/series/7651

Protein Folding, Misfolding, and Disease

Methods and Protocols

Edited by

Andrew F. Hill

Department of Biochemistry and Molecular Biology, Bio21 Molecular Science and Biotechnology Institute, University of Melbourne, Parkville, VIC, Australia

Kevin J. Barnham

Department of Pathology, Bio21 Molecular Science and Biotechnology Institute, University of Melbourne, Parkville, VIC, Australia

Stephen P. Bottomley

Department of Biochemistry and Molecular Biology, Monash University, Clayton, VIC, Australia

Roberto Cappai

Department of Pathology, Bio21 Molecular Science and Biotechnology Institute, University of Melbourne, Parkville, VIC, Australia

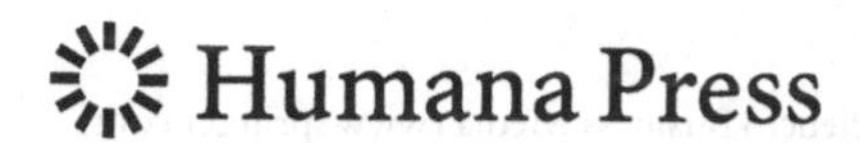

Editors
Andrew F. Hill
Department of Biochemistry
and Molecular Biology
Bio21 Molecular Science
and Biotechnology Institute
University of Melbourne
Parkville, VIC, Australia
a.hill@unimelb.edu.au

Kevin J. Barnham
Department of Pathology
Bio21 Molecular Science
and Biotechnology Institute
University of Melbourne
Parkville, VIC, Australia
kbarnham@unimelb.edu.au

Stephen P. Bottomley
Department of Biochemistry
and Molecular Biology
Monash University
Clayton, VIC, Australia
steve.bottomley@monash.edu

Roberto Cappai
Department of Pathology
Bio21 Molecular Science
and Biotechnology Institute
University of Melbourne
Parkville, VIC, Australia
r.cappai@unimelb.edu.au

ISSN 1064-3745 e-ISSN 1940-6029
ISBN 978-1-4939-5689-0 ISBN 978-1-60327-223-0 (eBook)
DOI 10.1007/978-1-60327-223-0
Springer New York Dordrecht Heidelberg London

Printed on acid-free paper

Humana Press is part of Springer Science+Business Media (www.springer.com)

Preface

Protein Folding, Misfolding, and Disease: Methods and Protocols brings together a collection of current methods for studying the analysis of protein folding and misfolding. Protein misfolding into structures, such as amyloid, is a key feature of many disorders of humans in which over 20 proteins are known to misfold and cause disease.

The protocols in this volume cover strategies for expressing and refolding recombinant proteins which can then be utilized in subsequent experiments. Methods for analyzing the formation of amyloid are given and are followed by detailed protocols for determining the size and structure of native and misfolded proteins. Specific examples of where misfolded proteins can be examined using state-of-the-art technologies are given in the final chapters in the book and serve as a reference for use with other protein systems. Together, the chapters in this volume provide an up to date collection of current methods in the area of protein folding and misfolding. The notes sections at the end of each of the chapters provide useful insights into the experimental techniques which will no doubt benefit researchers wanting to utilize these technologies.

We would like to thank the contributing authors for their efforts in contributing their chapters for this volume and also John Walker (Series Editor) for his patience and helpful advice in producing this book.

Parkville, VIC, Australia — ***Andrew F. Hill,***
Parkville, VIC, Australia — ***Kevin J. Barnham***
Clayton, VIC, Australia — ***Stephen P. Bottomley***
Parkville, VIC, Australia — ***Roberto Cappai***

Contents

Contributors

MARIE-ISABEL AGUILAR • *Department of Biochemistry and Molecular Biology, Monash University, Clayton, VIC, Australia*
STEPHEN P. BOTTOMLEY • *Department of Biochemistry and Molecular Biology, Monash University, Clayton, VIC, Australia*
LOUISE J. BROWN • *Department of Chemistry and Biomolecular Sciences, Macquarie University, Sydney, NSW, Australia*
ASHLEY M. BUCKLE • *Department of Biochemistry and Molecular Biology, Monash University, Clayton, VIC, Australia*
LISA D. CABRITA • *Department of Structural and Molecular Biology, University College London, London, UK; School of Crystallography, Birkbeck College, London, UK*
JOHN CHRISTODOULOU • *Department of Structural and Molecular Biology, University College London, London, UK; School of Crystallography, Birkbeck College, London, UK*
DAVID T. CLARKE • *Central Laser Facility, Science and Technology Facilities Council, Research Complex at Harwell, Rutherford Appleton Laboratory, Didcot, UK*
JAMES A. COOKE • *Department of Chemistry and Biomolecular Sciences, Macquarie University, Sydney, NSW, Australia*
GABRIELA A.N. CRESPI • *St. Vincent's Institute of Medical Research, Fitzroy, VIC, Australia*
GLYN L. DEVLIN • *Department of Chemistry, University of Cambridge, Cambridge, UK*
CHRISTOPHER M. DOBSON • *Department of Chemistry, University of Cambridge, Cambridge, UK*
JOHN D. GEHMAN • *School of Chemistry, Bio21 Molecular Science and Biotechnology Institute, The University of Melbourne, Parkville, VIC, Australia*
KENNETH N. GOLDIE • *Bio21 Molecular Science and Biotechnology Institute, The University of Melbourne, Parkville, VIC, Australia*
SALLY L. GRAS • *Department of Chemical and Biomolecular Engineering, Bio21 Molecular Science and Biotechnology Institute, The University of Melbourne, Parkville, VIC, Australia*
MICHAEL D.W. GRIFFIN • *Department of Biochemistry and Molecular Biology, Bio21 Molecular Science and Biotechnology Institute, The University of Melbourne, Parkville, VIC, Australia*
SEN HAN • *Department of Biochemistry and Molecular Biology and Department of Pathology, Bio21 Molecular Science and Biotechnology Institute, University of Melbourne, Parkville, VIC, Australia*
DANNY M. HATTERS • *Department of Biochemistry and Molecular Biology, Bio21 Molecular Science and Biotechnology Institute, The University of Melbourne, Parkville, VIC, Australia*

ANDREW F. HILL • *Department of Biochemistry and Molecular Biology, Bio21 Molecular Science and Biotechnology Institute, University of Melbourne, Parkville, VIC, Australia*
XU HOU • *Department of Biochemistry and Molecular Biology, Monash University, Clayton, VIC, Australia*
GEOFFREY J. HOWLETT • *Department of Biochemistry and Molecular Biology, Bio21 Molecular Science and Biotechnology Institute, The University of Melbourne, Parkville, VIC, Australia*
LIN WAI HUNG • *Department of Pathology, Bio21 Molecular Science and Biotechnology Institute, The University of Melbourne, Parkville, VIC, Australia*
JOHN A. KARAS • *Bio21 Molecular Science and Biotechnology Institute, The University of Melbourne, Parkville, VIC, Australia*
TUOMAS P.J. KNOWLES • *Department of Chemistry, University of Cambridge, Cambridge, UK*
ARIO DE MARCO • *University of Nova Gorica (UNG), Rožna Dolina (Nova Gorica), Slovenia.*
LUKE A. MILES • *St. Vincent's Institute of Medical Research, Fitzroy, VIC, Australia*
YEE-FOONG MOK • *Department of Biochemistry, Queen's University, Kingston, ON, Canada*
MICHAEL W. PARKER • *St. Vincent's Institute of Medical Research, Fitzroy, VIC, Australia*
KEYLA A. PEREZ • *Department of Pathology, Bio21 Molecular Science and Biotechnology Institute, The University of Melbourne, Parkville, VIC, Australia*
CHI LE LAN PHAM • *Department of Pathology, Bio21 Molecular Science and Biotechnology Institute, The University of Melbourne, Parkville, VIC, Australia*
JENNIFER PHAN • *Department of Biochemistry and Molecular Biology, Monash University, Clayton, VIC, Australia*
DENIS B. SCANLON • *Bio21 Molecular Science and Biotechnology Institute, The University of Melbourne, Parkville, VIC, Australia*
JASON SCHMIDBERGER • *Department of Biochemistry and Molecular Biology, Monash University, Clayton, VIC, Australia*
FRANCES SEPAROVIC • *School of Chemistry, Bio21 Molecular Science and Biotechnology Institute, The University of Melbourne, Parkville, VIC, Australia*
DAVID H. SMALL • *Menzies Research Institute, University of Tasmania, Hobart, TAS, Australia*
ADAM M. SQUIRES • *Department of Chemistry, The University of Reading, Reading, UK*
LYNNE J. WADDINGTON • *Division of Molecular and Health Technologies, CSIRO, Parkville, VIC, Australia*
ANDREW D. WATT • *Department of Pathology, Bio21 Molecular Science and Biotechnology Institute, The University of Melbourne, Parkville, VIC, Australia*
CHRISTOPHER A. WAUDBY • *Department of Structural and Molecular Biology, University College London, London, UK; School of Crystallography, Birkbeck College, London, UK*
MARK E. WELLAND • *Nanoscience Centre, University of Cambridge, Cambridge, UK*
NASRIN YAMOUT • *Department of Biochemistry and Molecular Biology, Monash University, Clayton, VIC, Australia*

Chapter 1

Strategies for Boosting the Accumulation of Correctly Folded Recombinant Proteins Expressed in *Escherichia coli*

Ario de Marco

Abstract

The yields of soluble recombinant protein expressed in bacteria can be significantly enhanced by optimally exploiting the cell-folding machinery. The proposed protocol describes the strategies that can be used to reach a suitable ratio between heat-shock proteins and target proteins. Specifically, molecular recombinant chaperones can be overexpressed or cell-native chaperones are stimulated by inducing chemical and physical stress. Furthermore, the protein synthesis block can make available the cell-folding machinery for in vivo, disaggregating and refolding the already produced misfolded recombinant target protein. A rapid fluorimetric analytical method allows the evaluation of the protein monodispersity at any single purification step and enables comparison of different growth combinations that are useful to test for screening the optimal conditions for each recombinant protein.

Key words: Bacterial recombinant expression, In vivo protein refolding, Protein aggregates, Aggregation index, Heat-shock proteins

1. Introduction

Protein folding is a critical step even under physiological conditions, but it may turn into a major limiting step during the overexpression of heterologous proteins. Cells must recognize between unfolded and aggregated proteins that can be recovered and converted into their native forms and other aggregates that must be degraded or stored in mostly inactive structures, like aggresomes and inclusion bodies (1). A complex network of proteins, like chaperones, foldases, and proteases supervise the dynamic exchange between insoluble precipitates, soluble aggregates, and correctly folded proteins (2). The reversible status of single proteins that can unfold, precipitate, and refold has been demonstrated in

Andrew F. Hill et al. (eds.), *Protein Folding, Misfolding, and Disease: Methods and Protocols*, Methods in Molecular Biology, vol. 752, DOI 10.1007/978-1-60327-223-0_1,

Escherichia coli using both wild-type bacterial proteins subjected to heat-shock and overexpressed recombinant proteins (3, 4).

Understanding the molecular processes supervising the cell's protein quality control machinery has aided two main biotechnological strategies: (a) aggregation prevention and (b) promoting in vivo refolding from aggregates. Both can be boosted, but each target protein has particular structural features, and therefore needs a specific approach in order to be produced in an active native form (5). Protein chains can also be stabilized by the fusion with opportune partners and their expression positively influenced by the constant availability of tRNAs and by increasing the half-life of their mRNA (6). Therefore, the small-scale empirical comparison among several expression conditions is beneficial for identifying the optimal expression protocol for that specific protein.

The following protocols detail the screening steps aimed at the selection of the culture combination that would yield the highest amount of soluble and monodispersed recombinant proteins starting from a minimal number of conditions. We chose two strains that improve the efficiency of protein translation, a sophisticated system based on molecular chaperone coexpression and two easy-to-realize alternatives aimed at boosting the stress-induced cell overexpression of heat-shock proteins. Particular emphasis is dedicated to the analysis of the recombinant protein produced.

2. Materials

2.1. Vectors and Strains

1. Expression vectors, pETM14 and pETM66, with selective kanamycin resistance (see Note 1).
2. BL21(DE3) bacterial cells (see Note 2). Rosetta (DE3) strain (chloramphenicol resistance, Novagen) enables the overexpression of tRNAs for rare codons. The mRNA is stabilized in the BL21 Star (DE3) strain (no resistance, Invitrogen) that is a deletion mutant for the RNase E.

2.2. Chaperone Overexpressing Cells

1. Molecular chaperone vectors, pBB540 (chloramphenicol), pBB542 (spectinomycin), and pBB572 (ampicillin) (see Note 3).
2. Stock solutions (1,000×) of the antibiotics chloramphenicol (10 mg/mL dissolved in ethanol), carbenicillin (100 mg/mL), kanamycin (30 mg/mL), and spectinomycin (50 mg/mL). Store them at –20°C.
3. Luria–Bertani (LB) medium
4. Prepare the competent cell (CC) buffer by adjusting to 6.7 with KOH, the pH of a solution containing 10 mM HEPES, 15 mM $CaCl_2$, and 250 mM KCl, and only later add 55 mM $MnCl_2$ to avoid precipitation (the solution turns brownish).

5. Chemical competent cells were prepared starting from a 500 mL *E. coli* culture grown in LB medium using a 2-L flask. The culture is grown at 30°C until the OD_{600} is 0.4. The bacteria are pelleted in a refrigerated centrifuge (10 min at 10,000×*g*) and gently resuspended on ice using 100 mL CC buffer prechilled at 4°C. The bacteria are then incubated for 20 min on ice and harvested as before. Finally, resuspend the bacteria in 18.6 mL CC+1.4 mL dimethyl sulfoxide (DMSO), incubate 20 min on ice, aliquot into Eppendorf tubes, freeze using liquid nitrogen, and store at −80°C.
6. Thermostatic orbital shaker with tube racks and bottle adaptors.
7. Chaperone overexpressing bacteria were obtained by transforming chemically competent wild type BL21(DE3) cells with the three vectors harboring the molecular chaperones. Stock solutions were prepared by inoculating 3 mL of LB medium in a 15-mL tube with a single bacterial colony. Shake the culture at 30°C until the OD_{600} reaches 0.4, and then add 600 μL of 80% glycerol to a 1.4 mL aliquot of the bacterial culture. Divide the stock into three aliquots, freeze them in liquid nitrogen, and store at −80°C. Induce the remaining bacteria with 0.2 mM IPTG, culture for a further 4 h, pellet the bacteria, and boil them in the presence of SDS-loading buffer to verify the chaperone overexpression after colloidal staining of proteins separated by SDS–PAGE (see Note 4).

   Using the frozen cell stock, inoculate 500 mL of LB and prepare chemically competent chaperone-overexpressing cells. Stocks remain competent for several months.

2.3. Cell Culture

1. LB medium and glycerol (70%).
2. 15-mL disposable tubes (see Note 5).
3. Autoinducible growth medium was prepared according to Studier (7).
4. Plastic inoculation tips (Sarstedt).
5. Chloramphenicol at a concentration of 200 mg/mL (stock solution), dissolved in ethanol, to block the protein synthesis.
6. Thermostatic orbital shaker with tube racks (New Brawnswick).
7. Bench centrifuge (Eppendorf).
8. Benzyl alcohol (Sigma): The product is toxic, and handle with care (see Note 6).

2.4. Analytical Protein Purification

1. Lysis buffer for IMAC: Contains 50 mM Tris–HCl, pH 8.0, 500 mM NaCl, 5 mM $MgCl_2$, and 1 mg/mL lysozyme.
2. DNase I (Roche), prepare stock solution, and store aliquots at −20°C.

3. Washing buffer: 50 mM Tris–HCl, pH 8.0, 500 mM NaCl, 15 mM imidazole, and 0.02% Triton X-100. Prepare stock solutions of Tris–HCl (1 M), NaCl (5 M), and imidazole (2 M). Conserve Triton X-100 in the dark.
4. Ni Sepharose beads are stored at 4°C in the presence of preservatives and must be washed at room temperature before use. Pipet 25 μL of Ni Sepharose beads into a 1.5-mL Eppendorf tube in the presence of 300 μL of PBS buffer. After careful mixing by several tube top/bottom invertions, the beads are separated by centrifugation (2 min at 3,300 × *g* in a bench minifuge) from the supernatant that is gently removed. The procedure is repeated twice and finally, 30 μL of fresh lysis buffer is added to cover the beads and protect them from drying.
5. SDS-sample buffer (2×): 100 mM Tris–HCl pH 6.8, 4% (v/v) SDS, 5 mM DTT, 20% (v/v) glycerol, and bromophenol blue (4.8 mg/100 mL). Aliquots can be stored at −20°C.
6. A wheel for rotating Eppendorf and Falcon tubes allows the uniform mixing of suspensions.
7. Colloidal blue dye (Instant Blue, Novexin) is more sensitive and rapid than conventional coomassie staining. Furthermore, it does not need gel fixation, and the washing is done in water (no methanol and acetic acid in the lab).
8. Bench spectrophotometer (Eppendorf).
9. Water bath sonicator for improving bacteria lysis.

2.5. Second Optimization Round

1. Lysis buffer for large-scale purifications: 50 mM Tris–HCl, pH 8.0, 500 mM NaCl, 10% (v/v) glycerol, 20 mM imidazole, 1 mM DTT. 5 mM $MgCl_2$, 50 μg/mL DNase I, and 1 mg/mL lysozyme are added after cell pellet has been completely resuspended.
2. Supernatant filtration (0.22 μm filter, Vivaspin) is strongly suggested before starting FPLC.
3. FPLC system: The affinity purification can also be performed without an FPLC system, but its use is suggested since it enables the setup and more accurate comparison of purification parameters. Purification can be performed at both 4°C and room temperature, depending on the properties of the protein.
4. Hi-Trap Chelating Sepharose column for IMAC (see Note 7): The column is stored in 20% ethanol at 4°C and must be first washed in 5-column volumes of water, then charged with the metal ions (20 mM stock solution), and finally washed in 5-column volumes of washing buffer before being loaded with the protein sample. The column should be pre-equilibrated at the purification temperature.

5. Desalting column (5 mL Hi-Trap Desalting – GE Healthcare): The column is stored in 20% ethanol at 4°C and must be washed with 5 volumes of water using a 25-mL syringe before being equilibrated by loading 5 volumes of the suitable buffer.

2.6. Biophysical Characterization of the Purified Proteins

1. Size Exclusion Chromatography (SEC) was performed using a Superose 12 10/300 GL column (GE Healthcare). The column is stored in 20% ethanol at 4°C and must be carefully washed in water initially using low flow rates (0.20–0.25 mL/min) to avoid overpressure due to the resin volume changes. Later, the flow rate can be set at 0.5 mL/min for optimal peak separation. Wash with 5-column volumes of water and equilibrate with 5-column volumes of SEC buffer (25 mM Tris–HCl, pH 8.0, 250 mM NaCl). The accuracy of SEC depends on the reliability of the pumps controlling the liquid flux through the column. An FPLC system is, therefore, strongly suggested.
2. Spectrofluorimeter (Jasco): The working temperature (25°C) of the device must be stable before starting the measurements. Therefore, remember to switch on the equipment, plus water bath, at least 30 min before starting the experiments. Fix the excitation wavelength at 280 nm and scan the emission range between 260 and 400 nm.
3. Cuvette for fluorimeter (cat. n. 105.250 QS, light pass 10 mm, Hellma).

3. Methods

The two-step cell culture protocol (Fig. 1) has been developed as a biotechnological application of the observation that the recombinant protein expression block resulted in a recovery in a soluble form of already aggregated polypeptides (4). The in vivo disaggregation is performed by the cell-folding machinery and becomes more efficient when the amount of molecular chaperones has been previously boosted by recombinant overexpression (5, 8) or by physical and chemical stress factors, such as benzyl alcohol addition and incubation at 42°C (9) (see Note 8).

A simple comparison of the yield of soluble proteins recovered using different expression conditions can be misleading. The biophysical analysis of the purified protein is crucial for evaluating the quality of the material. However, the equipment necessary for accurate analyses of folding may not be always available, and suitable proteins are limited in quantity. For this reason, we present here a simple and fast alternative to measure protein monodispersity for which only tiny amounts of proteins are required. 250 mL

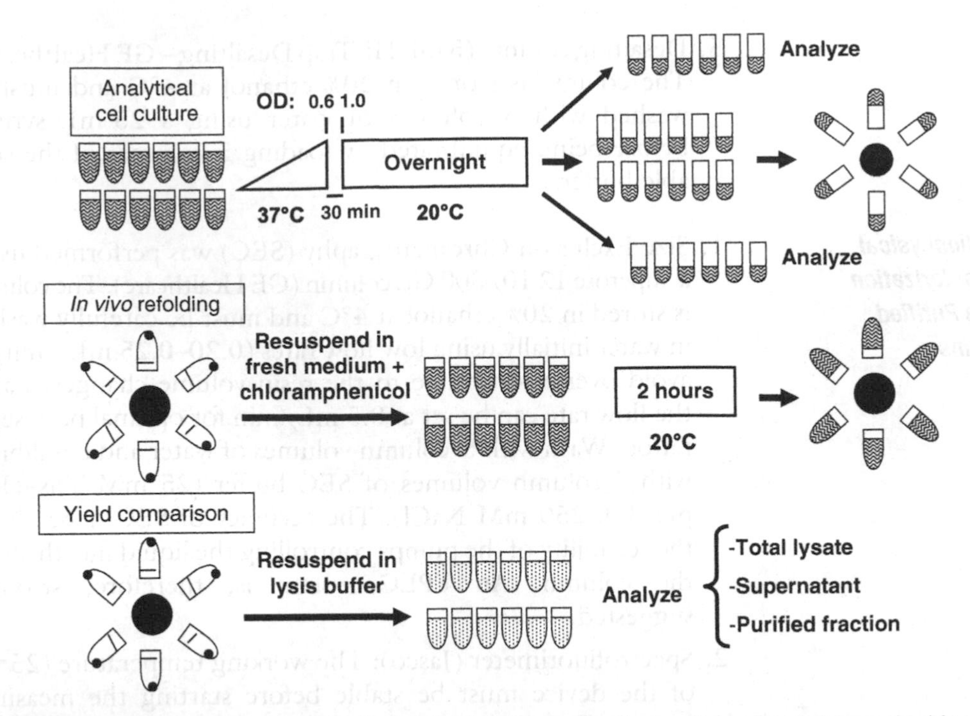

Fig. 1. Schematic representation of the analytical bacterial culture and purification. 3 mL of bacterial samples (six growth conditions × two different vectors) were initially grown using autoinducible media and 1.5 mL of each culture were pelleted and stored. The remaining cells were resuspended into fresh medium plus chloramphenicol at 200 μg/mL to block protein expression, cultured further for 2 h, and finally recovered by centrifugation. All the resulting 24 samples were lysated and affinity purified. Their fractions corresponding to the total lysate, soluble fraction, and IMAC-purified fraction were separated by SDS–PAGE and compared to evaluate the protein expression and final yields.

of culture medium of the few most promising conditions selected after the analytical screening is sufficient to purify enough proteins for folding qualitative comparison.

3.1. Comparative Small-Scale Protein Production for the Identification of Optimal Expression Conditions

1. The DNA sequence of interest is cloned into the two vectors, pETM14 and pETM66 (6). The corresponding polypeptides will result as fusions of the target protein, plus two different cassettes. The expression product of pETM14 will be a construct made of His-tag, recognition sequence for the 3 C protease, and target protein. With respect to the pETM14, the pETM66 introduces the NusA sequence at the N-term of the His-tag. NusA has been shown to stabilize passenger proteins and, in the screening analytical production, it is a model for the several solubilizing partner proteins, such as GST or MBP among others, proposed so far. A set of six different expression conditions for each of the two constructs will be compared.

2. Transform both vectors in competent bacteria from the strains, BL21(DE3), Rosetta (DE3), and BL21 Star (DE3), and in the chaperone overexpressing bacteria. Be careful in using Petri dishes with the right antibiotic resistances for plating the transformed bacteria, and incubate overnight at 37°C.

3. Use one single colony grown on the plates to inoculate 3 mL of autoinducible medium or LB medium (chaperone cotransformed bacteria) in a 15-mL plastic tube. For both vectors, inoculate three tubes with BL21(DE3) strain and one tube for each of the other strains. Insert the resulting 12 tubes in a rack inside a shaker thermostated at 37°C. A rack fixed at 30°C improves the culture aeration. However, at least 230 rpm would be necessary for the optimal bacterial growth.
4. Monitor the OD_{600} of the cultures. At values between 0.5 and 1, decrease the incubation temperature to 20°C, add 10 mM benzyl alcohol to one of the tubes transformed with BL21(DE3) before adding 0.2 mM IPTG to the chaperone overexpressing bacteria, and incubate another sample for 30 min in another shaker at 42°C. Both stress treatments induce the accumulation of heat-shock proteins. Let the cultures grow overnight at 20°C (see Note 9).
5. The successive day, recover 1.5 mL from each tube of the overnight cultures into an Eppendorf tube, centrifuge the 12 tubes for 3 min at 16,000 × *g*, discard the supernatant, and freeze the pellets.

3.2. Expression Conditions for Inducing In Vivo Protein Refolding

1. Centrifuge the remaining 1.5 mL of the overnight cultures directly in the 15-mL tubes (5 min × 3,300 × *g*) and discard the supernatant.
2. Gently resuspend all the pellets in fresh LB medium, plus 200 μg/mL chloramphenicol, using a micropipette (dilution 1:1,000 of the 200 mg/mL stock solution). Incubate the bacteria in a shaker (230 rpm) at 20°C for 2 h.
3. Transfer the bacteria culture in Eppendorf tubes and pellet them by centrifugation (3 min × 16,000 × *g*). Discard the medium and freeze the 12 pellets.

3.3. Small-Scale Protein Purification

1. Resuspend the pellets of the 24 samples in 400 μL of lysis buffer and incubate at room temperature under constant mixing using a rotating wheel set at a speed of 20 rpm.
2. Sonicate in a water bath for 5 min, then add 50 μg/mL DNase I to each tube, and let incubate for 20 min at the same conditions as above.
3. Remove 2.5 μL from each tube and mix them with 2.5 μL of water and 5 μL of the SDS-sample buffer. These samples from total lysates are separated by SDS–PAGE and are used to identify the level of expression of the target protein. Such information is useful to verify that the protein of interest is actually expressed in a specific growth combination and to estimate, by comparison with soluble and purified fractions, the percentage of the protein that are purified.

4. Centrifuge the tubes for 5 min at 16,000 × *g* and mix 5 μL of each supernatant with 5 μL of SDS-sample buffer. These samples represent the soluble fractions. Add the rest of the supernatant to the Eppendorf tubes in which the Ni Sepharose has already been washed and equilibrated (see Note 10).
5. Incubate the samples for 30 min under constant rotation (30 rpm) at room temperature.
6. Recover the beads by centrifugation (2 min at 3,300 × *g*) and discard the supernatant. Using 1,000 μL tips for supernatant removal can generate a partial mix of the beads due to aspiration and a consequent loss of materials. To avoid these perturbations, carefully remove the upper supernatant volume using a 1,000 μL pipette and then complete the work using a 100 μL tip.
7. Resuspend the beads in 500 μL of washing buffer, incubate and centrifuge using the same conditions as above, remove the supernatant, and repeat the complete washing procedure.
8. Add 30 μL of SDS-sample buffer to the beads recovered after the second washing step and boil them before running an SDS–PAGE. These samples represent the purified proteins. Unexpected differences in protein amount between soluble and purified samples can be observed (see Note 11).
9. Compare the yields obtained using the different growth conditions and select a couple of them for the second screening step (see Note 12).

3.4. Large-Scale IMAC Purification of Selected Protein Growth Conditions

1. For each of the selected conditions, inoculate 10 mL of LB medium plus antibiotics and 1% glycerol with the frozen cell stock and let grow overnight in a 50-mL flask at 30°C.
2. Use 2.5 mL of the overnight culture to inoculate 250 mL of autoinducible medium or LB medium (chaperone cotransformed bacteria) in a 1-L flask. Incubate at 37°C at 180 rpm until the bacterial growth becomes visible, then switch to 20°C, add 0.2 mM IPTG to the chaperone overexpressing bacteria, and leave to grow overnight. When benzyl alcohol must be added, wait until the OD_{600} has reached 1.
3. The day after, collect the pellet by centrifugation (20 min × 3,000 × *g*) following the opportune protocol (conventional or two steps with in vivo refolding) and resuspend it in 13 mL PBS. Transfer the suspension into a 15-mL Falcon tube, centrifuge for 15 min at 3,000 × *g*, discard the supernatant, and freeze the pellet.
4. Pellets (around 1.5 g) are resuspended by vortexing into 5 mL of lysis buffer, and cells are disrupted by sonication (5 min in a water bath). Add lysozyme, DNase, and $MgCl_2$, and incubate the samples for 30 min with constant agitation.

Neither nonlysed cells (cloudy materials) nor DNA (viscous material) should be visible.

5. Centrifuge the lysate (30 min × 120,000 × *g*), filter the supernatant through a 0.22 μm filter, and use it to load a pre-equilibrated His-Trap column connected to a suitable FPLC.
6. Wash the column thoroughly until no protein leaks out. The absorbance at 280 nm should be close to the baseline measured when the column was equilibrated with the washing buffer.
7. Elute the His-tagged protein with the elution buffer. Observe the absorbance value and begin to recover the corresponding sample in clean tubes.
8. High concentrations of imidazole can destabilize the protein and, therefore, it is advisable to immediately exchange the sample buffer. This step also provides the optimal conditions for the next step (protease digestion, chromatographic separation, or storage). Desalting columns are faster than dialysis and can be scaled-up to different volumes. The 5 mL Hi-Trap Desalting column allows the accomplishment of the operation in few minutes and can also be operated manually with modest dilution.

   Briefly, load 1.5 mL of protein sample on the top of the pre-equilibrated 5-mL column using a syringe. Pay attention to avoid air bubbles in the needle. The exact injected volume can be determined by collecting the liquid dropping out of the column bottom into a 1.5-mL Eppendorf. Successively, load the column with the buffer of choice and recover the protein into a 2-mL Eppendorf. This fraction contains the protein in the new buffer.

   Larger volumes can be handled by HiPrep 26/10 desalting columns mounted on an FPLC. In this case, the online detection of both conductivity and proteins is necessary to obtain an accurate separation of the protein from the salt.
9. Measure the aggregation index (AI, see next paragraph).

3.5. Protein Aggregation Analysis (*see* Note 13)

1. SEC is a chromatographic step that separates proteins depending on their mass and shape. Therefore, it can be used for both polishing the target protein from minor contaminants and discriminating monomers from polymers and large aggregates. Proteins should be concentrated, since only small volumes can be loaded and the chromatographic step dilutes the samples and their corresponding absorbance peaks.

   Inject 250 μL of protein (at a concentration of at least 1 mg/mL) onto the top of the pre-equilibrated column and record the absorbance at 280 nm. Recover the samples at the bottom of the column. In Fig. 2, the chromatograms of protein

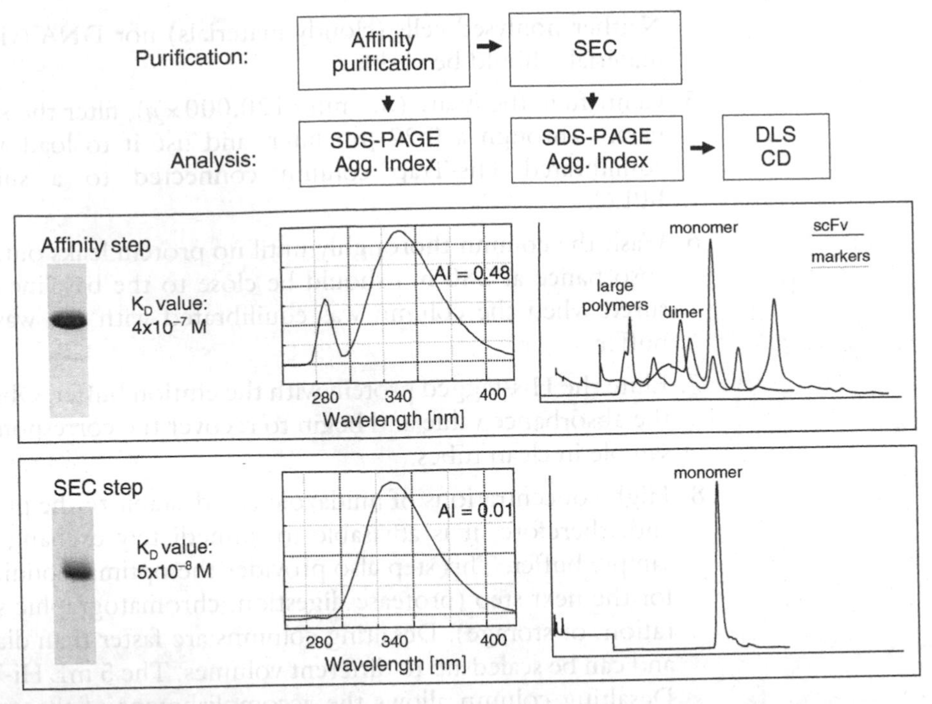

Fig. 2. Identification of protein-soluble aggregates. SDS–PAGE analysis must be coupled to AI determination to evaluate the quality of the purified protein at any purification step. Further structural information can be finally inferred by dynamic light scattering and circular dichroism analyses. In the reported example, an scFv antibody was purified to almost homogeneity after affinity purification, as detectable by SDS gel staining. However, SEC evidenced that the antibody was present in solution in different forms, from monomer to large aggregates. The monomer had an AI value significantly lower (0.01 instead of 0.48 of the post-affinity sample) and a higher affinity (K_D 5×10^{-8} instead of 4×10^{-7}) for the antigen, as measured by surface plasmon resonance.

standards and of a recombinant antibody sample are reported as an example (see Note 14).

2. Fluorimetric aggregation index (see Note 15): The equipment must be already set and equilibrated. Load 120 μL into the cuvette, check the sensitivity to select the suitable voltage, and operate the scan between 260 and 400 nm. Register the AI and compare it with the AI values of the samples expressed using different expression conditions and after each purification step.

4. Notes

1. Parallel cloning, by cut-and-paste the same original PCR product in a conserved multicloning site shared by several expression vectors that differ only for the fusion partners, is one of the options to compare the expression (total and soluble yields) of the constructs. The pETM vectors always have a His-tag and

a proteolytic cleavage site (as in the case of pETM14). Moreover, they can provide a second, larger stabilizing tag (such as the NusA of pETM66). A vast variety of carriers are available (thioredoxin, GST, MBP, DsbC, DsbA, GB-domain, Z-domain) (6). Similar collections have been produced compatible for ligation-independent cloning (10) and recombination-based systems (11). The same principle can be used to generate application-ready vectors by fusing sequences that are not intended to stabilize the target protein, but to directly enable its use in downstream experiments (fusion to GFP, to biotinylated sequences, and to antibody-detectable tags).

2. The strain BL21(DE3) is *lon*⁻ and *ClpA*⁻ and, consequently, its proteolytic activity is negligible and the addition of protease inhibitors is not necessary. Furthermore, the T7 promoter can be exploited in combination with the autoinducible medium (7) to promote uniform protein expression in the absence of time-consuming OD measurements of the cell culture and IPTG addition. Usually, both total cell and protein yields are higher when bacteria are cultured in autoinducible medium rather than in LB medium. However, the plasmid harboring the chaperone constructs is not suitable for autoinducible medium.
3. The vectors have been developed in the lab of Professor B. Bukau – University of Heidelberg – and are available on request at the author's address: Ario de Marco, *University of Nova Gorica (UNG), Vipayska 13, PO Box 301-SI-5000, Rožna Dolina (Nova Gorica), Slovenia.ario.demarco@ung.si.*
4. Cotransformation with three different plasmids can be difficult when cell competence is not optimal. We typically use at least 50 ng for each plasmid. In the worst case, you can transform with a single plasmid each time and then prepare competent cells for as many steps as necessary.

The coexpression of several recombinant proteins can be for itself a metabolic burden for the host cell. The total expression level of the target protein is usually significantly lowered, but compensated by a higher ratio between soluble and insoluble proteins. The main limitation of the approach is that because of the necessity to work with four different antibiotic resistances, there is no possibility to cotransform with, for instance, a plasmid for the overexpression of rare-codon tRNAs. Similarly, several combinations of protein solubility effectors are prevented by the lack of sufficient selective markers. Nevertheless, it has been observed that the expression leakage is prevented in cells hosting several plasmids (12). Carbenicillin is preferable to ampicillin in long-term culture to avoid lactamase activity that completely digests the antibiotic and, consequently, removes the selective pressure necessary to maintain the corresponding vector.

Some proteins are stabilized only by the presence of partners with which they form complexes in vivo. The coexpression of complex subunits by using a polycistronic vector has been successfully performed in combination with the overexpression of molecular chaperones (5).

5. The protocol reported in this page considers a standard bench approach. Nevertheless, microplates can be used for culturing and handling large numbers of protein-solubility effector combinations resulting from mutually combining growth conditions, cell cultures, bacteria strains, vectors with different fusion partners, osmolytes, and molecular chaperones.
6. The benzyl alcohol mechanism of action is not thoroughly understood, but it has been shown that its addition to bacteria and mammalian cells results in a coordinated activation of heat-shock genes (13).
7. Different metal ions have been proposed for IMAC. For instance, IMAC of total lysates with Fe^{3+} resulted in phosphoprotein enrichment, but $NiCl_2$ and $MnCl_2$ are the most commonly employed salts for the purification of His-tagged proteins. In our tests, the use of $NiCl_2$ correlated with the highest yields, but with significant nonspecific background. In contrast, purifications with Mn^{2+} yielded highly pure target protein with negligible presence of contaminants.
8. Both NaCl and ethanol have been used to trigger a heat-shock-like response in cultured cells (8). However, these stress factors can directly induce aggregation, and a setup of the conditions would be necessary to obtain a net beneficial effect.
9. Low growth temperature slows down the bacteria growth, and recombinant proteins find lower competition for the cell-folding machinery. As a consequence, longer incubation times are necessary, but the yields of soluble recombinant proteins are higher. Therefore, *E. coli* is mostly cultured overnight at 16–20°C.
10. The same strategy can be used with GST-fusion proteins and glutathione-affinity purification. For more details, see the protocol published in (8).

    Sepharose beads can be substituted by magnetic beads. This alternative is more expensive, but its main advantages are (a) to avoid the centrifugation steps, substituted by faster magnetic separation and (b) better separation between beads and liquid phase, with improved washing efficacy. The protocol that uses magnetic beads can easily be adapted to a robot washer to automate the complete cycle.
11. IMAC can fail to purify correctly folded proteins mainly because of poor accessibility of the His-tag. This can be trapped between the passenger and target protein or simply

folded inside its globular structure. Furthermore, several soluble aggregates do not bind to affinity resins, even though the His-tag accessibility cannot be used as a discriminating parameter due to its ease of being recognized even in misfolded proteins.

12. The protein amount that would be possible to recover by imidazole elution at the end of the small-scale purification would be often sufficient to measure the AI at the fluorimeter. However, imidazole strongly interferes with the measurement and, consequently, must be removed. Microdialysis is a cumbersome and expensive procedure and, therefore, we suggest testing the protein monodispersity directly after the large-scale purification.

13. Dynamic light scattering is another method to evaluate the dispersity of proteins in solution, while circular dichroism provides obtain information concerning secondary and tertiary structures. Both methods need high protein concentrations, specific biophysical competences, and dedicated devices, the availability of which can be limited in nonspecialized labs.

14. The protein recovered after affinity chromatographic step is often highly pure. Nevertheless, polymerization/aggregation is often an issue. For instance, recombinant antibodies are prone to form increasingly larger polymers in the absence of salt (14), and the K_D of the dimers is significantly lower than that of the monomer (Fig. 2).

15. The AI has been introduced to identify and evaluate the incidence of the protein-soluble aggregates (15). Soluble aggregates are not visible, despite their often large mass (16), and therefore their presence is often missed which has consequences for the reliability of the purification and downstream uses of the protein (Fig. 3). AI is the ratio between the fluorescence emission measured at 280 and 340 nm following the excitation at 280 nm. The signal at 280 nm is due to light scatter and increases with the aggregation of the sample while the emission at 340 nm is directly correlated to the accessibility of the aromatic groups in proteins. When they are masked by aggregation, the emission is low while monodispersed samples emit at high efficiency. The AI correlates well with SEC data (Fig. 2), which is fast to obtain even with dilute samples, and the material can be recovered. AI does not give absolute values, but the data are particularly useful for the comparison of production conditions and purification protocols, since an AI variation indicates the introduction of a condition that changes the protein monodispersity.

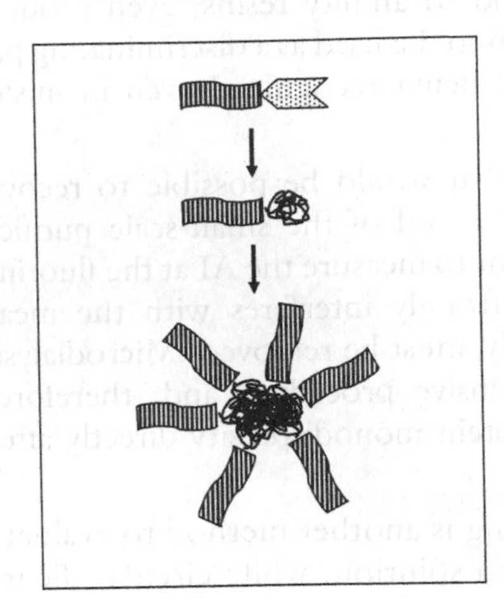

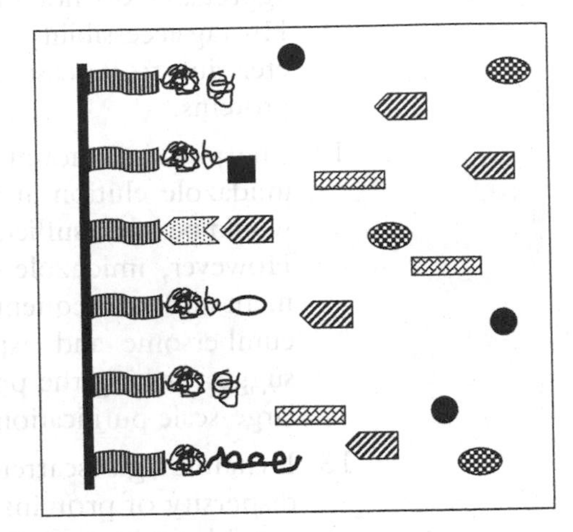

Fig. 3. Nature and shortcomings due to protein-soluble aggregates. Soluble aggregates often form when fusion proteins are used. The solubilizing partner is very stable, but the target protein misfolds. The misfolded moieties aggregate together, forming the core of micelles kept in solution by the soluble fusion partners exposed outwards (15). Typically, false results are generated when soluble aggregates of fusion proteins are used in pull-down experiments. The affinity tag moiety is efficiently bound by its specific resin, but most of the target proteins have not conserved a structure compatible with the specific binding of physiological partners. In contrast, exposed hydrophobic patches of the misfolded protein nonspecifically bind partners with other hydrophobic regions.

Acknowledgments

The author would like to thank B. Bukau, E. Deuerling, P. Goloubinoff, and G. Travé for the stimulating discussions and S. Bossi, D. Ami, and P. Capasso for technical assistance.

References

1. Kopito, R.R. (2000) Aggresomes, inclusion bodies and protein aggregation. *Trends Cell Biol.* **10**, 524–530.
2. Mogk, A., Tomoyasu, T., Goloubinoff, P., Rüdiger, S., Röder, D., Langen, H., and Bukau, B. (1999) Identification of thermolabile *E. coli* proteins: prevention and reversion of aggregation by DnaK and ClpB. *EMBO J.* **18**, 6934–6949.
3. Goloubinoff, P., Mogk, A., Ben Zvi, A. P., Tomoyasu, T., and Bukau B. (1999) Sequential mechanism of solubilization and refolding of stable protein aggregates by a bichaperone network. *Proc. Natl. Acad. Sci. USA* **96,** 13732–13737.
4. Carrió, M. M., and Villaverde, A. (2001) Protein aggregation as bacterial inclusion bodies is reversible. *FEBS Lett.* **489,** 29–33.
5. de Marco, A., Deuerling, E., Mogk, A., Tomoyasu, T., and Bukau, B. (2007) Chaperone-based procedure to increase yields of soluble recombinant proteins produced in *E. coli. BMC Biotechnol.* 7, 32.
6. Dümmler, A., Lawrence, A.-M., and de Marco, A. (2005) Simplified screening for the detection of soluble constructs expressed in *E. coli* using a modular set of vectors. *Microbial Cell Factories* **4**, 34.

7. Studier, F. W. (2005) Protein production by auto-induction in high-density shaking cultures. *Protein Expr. Purif.* **41**, 207–234.
8. de Marco, A. (2007) Protocol for preparing proteins with improved solubility by co-expressing with molecular chaperones in *Escherichia coli. Nat Protocols* **2**, 2632–2639.
9. de Marco, A., Vigh, L., Diamant, S., and Goloubinoff, P. (2005) Native folding of aggregation-prone recombinant proteins in *Escherichia coli* by osmolytes, plasmid- or benzyl alcohol-overexpressed molecular chaperones. *Cell Stress Chaperones* **10**, 329–339.
10. Cabrita, L. D., Dai, W., and Bottomley, S. P. (2006) A family of *E. coli* expression vectors for laboratory scale and high throughput soluble protein production. *BMC Biotechnol.* **6**, 12.
11. Nallamsetty, S., Austin, B. P., Penrose, K. J., and Waugh, D. S. (2005) Gateway vectors for the production of combinatorially-tagged His6-MBP fusion proteins in the cytoplasm and periplasm of *Escherichia coli. Protein Sci.* **14**, 1435–1442.
12. de Marco, A., and De Marco, V. (2004) Bacteria co-transformed with recombinant proteins and chaperones cloned in independent plasmids are suitable for expression tuning, *J. Biotechnol.* **109**, 45–52.
13. Balogh, G., Horvath, I., Nagy, E., Hoyk, Z., Benko, S., Bensaude, O., and Vigh, L. (2005) The hyperfluidization of mammalian cell membranes acts as a signal to initiate the heat shock protein response. *FEBS J.* **272**, 6077–6086.
14. Olichon, A., Schweizer, D., Muyldermans, S., and de Marco, A. (2007) Heating as a rapid purification method for recovering correctly-folded thermotolerant VH and VHH domains. *BMC Biotechnol.* 7, 7.
15. Nominé, Y., Ristriani, T., Laurent, C., Lefevre, J.-F., Weiss, E., and Travé, G. (2001) A strategy for optimizing the monodispersity of fusion proteins: application to purification of recombinant HPV E6 oncoprotein. *Prot. Engineer.* **14**, 297–305.
16. Nataliello, A., Santarella, R., Doglia, S. M., and de Marco, A. (2008) Physical and chemical perturbations induce the formation of protein aggregates with different structural features. *Protein Expr. Purif.* **58**, 356–361.

Chapter 2

An *Escherichia coli* Cell-Free System for Recombinant Protein Synthesis on a Milligram Scale

Luke A. Miles, Gabriela A.N. Crespi, Sen Han, Andrew F. Hill, and Michael W. Parker

Abstract

In vitro translation systems derived from a wide range of organisms have been described in the literature and are widely used in biomedical research laboratories. Perhaps the most robust and efficient of these cell-free systems is that derived from *Escherichia coli*. Over the past decade or so, experimental strategies have been developed which have enhanced the efficiency and stability of *E. coli* cell-free systems such that we can now prepare recombinant proteins on a scale suitable for purification and analysis by biophysical and structural biology techniques, which commonly require relatively large quantities of protein. This chapter describes in detail the protocols employed in our laboratory to prepare translationally active *E. coli* extracts and to synthesise proteins on a milligram scale from these extracts.

Key words: Cell-free protein synthesis, In vitro translation, *Escherichia coli*, S30

1. Introduction

Escherichia coli is a robust and widely used organism for the expression of recombinant proteins. In vivo expression in *E. coli* requires the gene of interest to be cloned into an appropriate expression vector, which is transformed into competent laboratory strains of *E. coli*. These transformed cells are grown in suitable media under antibiotic selection for the plasmid vector. Transcription or RNA synthesis is induced at mid-log phase of cell growth (for example, by the addition of isopropyl β-D-1-thiogalactopyranoside, IPTG), and the incubation is continued for several hours to enable translation of the gene product. *E. coli* cells are then harvested, lysed and the recombinant protein is purified from soluble fractions or from inclusion bodies.

Andrew F. Hill et al. (eds.), *Protein Folding, Misfolding, and Disease: Methods and Protocols*, Methods in Molecular Biology, vol. 752, DOI 10.1007/978-1-60327-223-0_2,

In cell-free protein synthesis, *E. coli* cells harbouring no expression plasmid are grown to mid-log phase, harvested, washed, and carefully lysed. The resultant suspension when clarified by centrifugation and dialysis contains active biomolecules that constitute the translational machinery of *E. coli*, including ribosomes, protein translation factors, aminoacyl tRNA synthetases and transfer RNA species. When coupled with an appropriately structured transcript of the desired gene (or DNA and an in vitro transcription system), an ATP regeneration system and other reagents, such as amino acids, these extracts can continue to efficiently synthesise proteins for many hours in a plastic tube, 96-well plate or dialysis apparatus.

In this chapter, we give a detailed account of how to prepare translationally active extracts of *E. coli* suitable for in vitro protein synthesis. Extracts are prepared several times a year in our laboratory, typically prepared from 6 L batches of *E. coli* culture, which will yield approximately 10 mL of extract, sufficient for approximately 40 mL of cell-free reactions. Once prepared, these extracts called S30 extracts can be stored at −80°C with little or no loss in activity for at least 12 months. In fact, we have successfully expressed proteins from extracts prepared 3 or 4 years prior with no discernible loss in efficiency. So while it may seem daunting to undertake the protocols described herein, once you have perfected the technique, enough extracts for a year's work can be prepared within a few weeks and frozen.

In addition to describing the preparation of S30 extracts, we also describe how to perform cell-free protein synthesis reactions, and the preparation and storage of stock reagent mixtures and buffers that enable cell-free reactions to be performed routinely. Cell-free reactions can be performed in batch mode for short-lived protein synthesis suitable for optimising reaction conditions and for screening expression levels and solubility of a given construct. For scaled up protein production, these reactions can be coupled with a dialysis system for ongoing supply of reactants, such as amino acids and ATP, and for the removal of by-products that may be inhibitory to translation. This dialysis mode can sustain protein synthesis over 8–24 h depending on reaction temperature. We have prepared as much as 5 mg of purified protein from a 1 mL dialysis mode reaction. This corresponds roughly to 33 mg of protein per 1 L of starting *E. coli* culture, which is equivalent to efficiencies expected for in vivo expression of recombinant proteins in *E. coli*.

Cell-free protein synthesis is an alternative to in vivo expression for highly efficient incorporation of labelled amino acids to aid structural studies. This includes ^{15}N, ^{13}C, and ^{2}H labelling of proteins for Nuclear Magnetic Resonance (NMR) spectroscopy studies and selenomethionine labelling of proteins for X-ray crystallographic studies of protein structure. We also routinely use cell-free synthesis to prepare troublesome proteins, including toxic proteins, integral membrane proteins, and those prone to misfolding, aggregation, and proteolysis. We have given a comprehensive overview of *E. coli* cell-free expression elsewhere (1), and there are many reports in the literature describing cell-free protein synthesis systems (2–7) that should be consulted for a full understanding of cell-free protein synthesis. Here, we provide step-by-step protocols used in our laboratory for *E. coli* extract preparation and for performing cell-free protein synthesis reactions.

2. Materials

2.1. E. coli Culture

1. *Luria–Bertani Broth* (LB) for *E. coli* starter cultures and expression plasmid amplification: Dissolve bacto-tryptone, sodium chloride, and yeast extract in distilled water at 10, 10, and 5 g/L, respectively. Adjust the pH to 7.0 with 5N NaOH. Autoclave and store at 4°C until required.
2. *Incomplete Rich Media* (IRM, see Note 1): Dissolve 5.6 g of KH_2PO_4, 28.6 g of K_2HPO_4 and 1 g of Yeast extract per litre of distilled water. After complete mixing, autoclave in 1 L batches and store at 4°C in 2 L baffled conical flasks capped with aluminium foil fastened with autoclave tape until required.
3. *IRM Supplement*: Combine 9.4 mg of thiamine, 125 g of d-glucose, and 62.5 mL of 0.1 M magnesium acetate stock solution and make up to 500 mL with distilled water. This supplement is made fresh before use and sterile filtered (0.22 μm).
4. *E. coli* glycerol stock: Commercial glycerol stocks of *E. coli*, such as BL21 and BL21 Star strains (Invitrogen).

2.2. S30 Extract Preparation and Cell-Free Reaction Reagents

Stock Solutions (see Note 2)

1. *Diethylpyrocarbonate (DEPC, 0.1%) treated milli-Q water*: DEPC is toxic on inhalation and is also an irritant to the eyes and mucous membranes. Skin contact should be avoided, and it should be handled in a fume cabinet. DEPC is added to 0.1% v/v to milli-Q water and the suspension mixed with a magnetic stirrer at 4°C overnight or longer, then sterilised by autoclave. A sickly sweet smell persists in autoclaved DEPC-treated water as hydrolysis of DEPC produces ethanol which in turn reacts with trace carboxylic acid to form volatile esters.

DEPC-Treated Stock Solutions (Made with Milli-Q Water then DEPC Treated).

2. Potassium acetate (6 and 5 M stocks of 500 mL, stored at 4°C).
3. Magnesium acetate (3 M, 500 mL, stored at 4°C).
4. Magnesium acetate (1.4 and 0.1 M stocks of 50 mL, stored at 4°C).
5. Polyethylene glycol 8000 (PEG8000, 50% w/v, 25 mL stored at room temperature).

Stock Solutions Made with DEPC-Treated Milli-Q Water and Sterile Filtered.

6. Tris–acetate (100 mM, pH 7.0, 1 L, stored at 4°C).
7. Tris–acetate (2.2 M, pH 8.2, 500 mL, stored at 4°C).
8. HEPES–potassium hydroxide (2 M, 500 mL, stored at 4°C).
9. Dithiothreitol (DTT) (0.55 M, 1–10 mL aliquots, stored at −20°C).
10. DTT (1 M, 1–10 mL aliquots, stored at −20°C).
11. Ammonium acetate (1.4 M, 1–10 mL aliquots, stored at −20°C).
12. Mixture of the 20 natural amino acids (25 mM each, 1–10 mL aliquots, stored at −20°C).
13. Creatine phosphate (1 M, 0.2 mL aliquots, stored at −20°C).
14. Creatine kinase (10 mg/mL in 10 mM Tris–acetate pH 7.0, Roche Diagnostics, Basel, Switzerland) stored at −80°C in 50 μL aliquots to avoid freeze–thaw cycles.
15. Cyclic adenosine monophosphate (cAMP, 50 mM, 1 mL aliquots, stored at −20°C).
16. Folinic acid (30 mM, 0.2 mL aliquots, store at −20°C).
17. *S30 Buffer*: This buffer is used for dialysis of extracts and should be made fresh during extract preparation to the required volume and cooled to 4°C; 14 mM magnesium acetate, 60 mM potassium acetate, 1 mM DTT, and 10 mM Tris–acetate (pH 7.0).
18. *Extract Incubation Mixture*: Made immediately prior to use in extract preparation; 4.4 mM DTT, 40 μM of all 20 amino acids, 293 mM Tris–acetate (pH 8.2), 13.2 mM adenosine triphosphate (ATP, from ~200 mM preparation, pH adjusted to 7.0), 9.2 mM magnesium acetate, 84 mM phosphoenol pyruvate (from ~0.42 M preparation, pH adjusted to 7.0), and 40 units of pyruvate kinase (from rabbit muscle, Sigma). Make mixture up to 6 mL with DEPC-treated water.

19. *Cell-Free Reaction Supplement*: This mixture is prepared in approximately 10–20 mL lots and stored long term at −20°C in small aliquots (0.5 mL) until required to prevent freeze–thaw cycles. 1 M potassium acetate (from 5 M stock); 6.8 mM DTT (from 0.55 M stock); 3.2 mM of each of uridine triphosphate (UTP), guanosine triphosphate (GTP), and cytidine triphosphate (CTP), all made fresh at stock concentrations of ~200 mM, pH adjusted to 7.0; 5 mM ATP (from ~200 mM stock made fresh, pH adjusted to 7.0); 2.56 mM cAMP (from 50 mM stock); 250 μM folinic acid (from 30 mM stock); 0.7 mg/mL *E. coli* tRNA (made fresh in 10 mM Tris–acetate buffer pH 7.0, Roche Diagnostics, Basel, Switzerland); 210 mM HEPES–KOH (from 2 M stock); 160 mg/mL final concentration of PEG8000 (from 50% w/v stock); and made to volume with DEPC-treated milli-Q water.
20. *T7 RNA polymerase*: Commercially available or prepared in-house by expression in *E. coli* and purified. This enzyme is stored at −80°C. The concentration of RNA polymerase is optimised for each cell-free expression system, but a good starting concentration is 70 μg of polymerase per 1 mL reaction depending on the activity of the sample. We store recombinantly expressed and purified T7 RNA polymerase at around 11 mg/mL.
21. *β-mercaptoethanol* (98%, Sigma).
22. *Dialysis tubing*.

    Dialysis tubing, Spectrapore 6000–8000 MWCO membrane with a 32 mm diameter (Spectrum Medical, Los Angeles, CA, USA) is prepared by soaking in DEPC-treated water overnight, then soaked for an hour in S30 buffer before use (all at 4°C).

3. Methods

3.1. S30 Extract Preparation

3.1.1. Day 1

1. Autoclave: 6 × 10 mL of *LB-broth* in 50-mL Falcon tubes (no antibiotic); 7 L of DEPC-treated milliQ water in 1-L laboratory bottles, each with a magnetic stirrer bar in place; 6 × 1 L of *IRM* in 2-L baffled flasks; 2 × 1-L glass measuring cylinders with magnetic stirrer bars and capped with aluminium foil; likewise, 3 × 2-L glass beakers each containing a heavy magnetic stirrer bar and foil capped.
2. While autoclave is in operation, prepare 500 mL of *IRM-Supplement* and sterile filter, cool to 4°C overnight.
3. Make 5 L of *S30 Buffer* from cooled DEPC-treated water (step 1) using measuring cylinders and bottles used in DEPC treatment, keeping magnetic stirrer bars in place.

4. Autoclave S30 buffer and cool to 4°C overnight.
5. Allow autoclaved IRM to stand at room temperature overnight.
6. Choose an *E. coli* cell line and innoculate the 6×10 mL LB-broth batches prepared in step 1 – incubate cultures overnight at 37°C with vigorous shaking.

3.1.2. Day 2

7. Add 80 mL of *IRM-Supplement* to each 1 L of *IRM* and immediately innoculate each 1 L of media with 10 mL (1/100 dilution) of overnight culture.
8. Incubate cultures at 37°C with vigorous shaking (150 rpm) until mid log-phase (0.8 OD_{600nm} as measured in an ultraviolet spectrophotometer). This should take approximately 4 h.
9. Harvest *E. coli* by centrifugation in 1-L buckets (5,000×*g*, 20 min, 4°C).
10. While spinning cultures, take 900 mL of cold S30 buffer and add 900 μL of 1 M DTT and 0.45 μL of β-mercaptoethanol and keep on ice.
11. Take up all pellets in a total of 300 mL of ice-cold S30 buffer (from previous step) and resuspend *E. coli* with repeated agitation with a 10 or 20-mL electric pipetter or similar device while on ice. Once a fine suspension is achieved, transfer suspension to 2×250-mL buckets, and pellet *E. coli* by centrifugation (5,000×*g*, 20 min, 4°C).
12. Repeat this wash step twice more until all 900 mL of the chilled reducing S30 buffer (step 10) is used. After the final wash step, discard supernatant and thoroughly drain buffer away from *E. coli* pellets by inverting tube onto paper towels.
13. At some stage during the wash steps or afterwards, warm 400 mL of S30 buffer (no DTT) in one of the glass beakers autoclaved on Day 1 and very slowly add 400 g of PEG8000 with constant stirring until a very fine white suspension is achieved. Then, autoclave the PEG8000/S30 buffer which should result in a clear viscous mixture when cooled.
14. Lysis of the *E. coli* is achieved by French Press or similar apparatus (such as a cell crusher) early on Day 3. Ideally, the apparatus should be washed repeatedly in DEPC-treated water and chilled overnight. Alternatively, the apparatus can be washed with ice-cold DEPC-treated water and allowed to stand to bring the temperature down immediately before lysis. Likewise, dialysis tubing should be soaked overnight for the following days work as described in Subheading 2.2.

3.1.3. Day 3

15. Prepare stock solutions for *Extract Incubation Mixture* as described in Subheading 2.2.
16. Thaw the cell pellets prepared on Day 2 on ice over a period of 1 h. While these thaw, place the PEG8000/S30 mixture, the two remaining sterile beakers, and four clean 50-mL centrifuge tubes into a refrigerator at 4°C.
17. Add 400 μL of 1 M DTT to the 50% w/v PEG8000/S30 mixture and stir in a cold room.
18. Take 100 mL of S30 buffer, add 100 μL of 1 M DTT, 5 μL of β-mercaptoethanol and keep covered on ice.
19. Complete wash of 6–8000 MWCO dialysis tubing in a small volume of cold S30 buffer and keep covered at 4°C.
20. Once the *E. coli* pellets have thawed, resuspend bacteria in the S30/DTT/β-mercoptoethanol buffer (step 18), and transfer to a 250-mL centrifuge bucket and pellet by centrifugation (5,000 × *g*, 20 min, 4°C).
21. During this centrifugation, transfer 11.2 mL of S30 buffer to a 50-mL Falcon tube and add 11.2 μL of 1 M DTT – keep covered on ice. This volume is calculated on the basis of a 9 g cell pellet and should be adjusted proportionally to the actual mass of *E. coli* obtained.
22. Prepare the cold French Press or cell crusher for loading with a final wash with ice-cold S30 buffer.
23. Resuspend pellet in the 11.2 mL S30/DTT buffer (step 21) and add it to the French press chamber and 15 μL of 1 M DTT to an ice-cold 50-mL centrifuge tube sitting on ice ready to receive lysate.
24. Press *E. coli* suspension into the DTT containing 50-mL centrifuge tube (on ice) at 8,400 psi until complete.
25. This lysate (~15 mL) is then subjected to centrifugation (30,000 × *g*, 30 min, 4°C) hence the name S30. Around 12 mL of cleared lysate should be obtained which should be carefully decanted away from cell debris into a second chilled centrifuge tube.
26. Repeat centrifugation on cleared lysate (30,000 × *g*, 30 min, 4°C). This should yield approximately 9 mL of supernatant.
27. During this second 30,000 × *g* spin, prepare the *Extract Incubation Mixture* as described in Subheading 2.2.
28. Carefully decant cleared lysate obtained from step 26 into a 50-mL Falcon tube and combine with 2.7 mL of freshly prepared *Extract Incubation Mixture*. This mixture is then incubated at 37°C with gentle swirling (preferably in a water bath) for 80 min. This step enables translation and degradation of transcripts to minimise background translation of endogenous

E. coli proteins. It also promotes precipitation of membranes and remaining cell debris.

29. During the pre-incubation step, transfer 1 L of cold S30 buffer to a sterile glass beaker, add 1 mL of 1 M DTT, and keep cool (4°C cold room on magnetic stirrer is suitable).
30. Dialyse incubated lysate against this buffer with stirring for 45 min in a cold room.
31. Replace with fresh buffer after 45 min and repeat dialysis twice more for a total of three dialysis steps against S30/DTT.
32. Transfer the dialysis membrane to the PEG8000/S30/DTT mixture and dialyse (concentrate) for 1 h with stirring in a cold room.
33. During this time, transfer 1 L of cold S30 buffer to the remaining sterile beaker, and add 1 mL of 1 M DTT. Once step 32 is completed, dialyse for another hour with stirring against this buffer at 4°C.
34. Remove extract from dialysis tubing with a needle and syringe and freeze in liquid nitrogen in small aliquots (170–1,000 μL) and store at −80°C until required.

3.2. Cell-Free Protein Synthesis Reactions

3.2.1. DNA Plasmid Preparation

Any T7-based plasmid vector suitable for in vivo expression in BL21 derived *E. coli* strains can be used in our cell-free expression system. In cell-free systems, unlike in vivo expression, plasmid amplification is performed separately from the culturing of *E. coli* for synthesis of recombinant protein. In vitro transcription/translation reactions are sensitive to contamination by salts, RNases, detergents, and alcohol, all of which are typically used in commercial plasmid purification kits. RNase often provided separately in plasmid purification kits should be excluded from the standard protocols when using these kits. A rule of thumb is the larger the scale of DNA preparation the better. Ideally, plasmids prepared by resin-based purification protocols should be further purified by phenol/chloroform and chloroform extraction. DNA should be precipitated with isopropanol, washed in ethanol, dried and taken up in RNase-free water to around 1 mg/mL. In our laboratory, we further purify plasmid preparations by dialysing overnight against S30 buffer at 4°C and store frozen in small aliquots.

Transcription reactions can be performed separately from translation according to instructions of the supplier of the bacteriophage RNA polymerase. We always perform coupled transcription/translation reactions with T7 RNA polymerase as the transcription enzyme. Mg^{2+}, DNA template, and RNA polymerase concentrations are the first parameters to be varied and optimised for each cell-free reaction system.

3.2.2. Performing Cell-Free Reactions

Having prepared and suitably stored the reagents and *E. coli* extract as described above, performing cell-free reactions is straightforward (see Note 3).

1. Place the following reagents described in Subheading 2.2 (reagent number shown in parenthesis) on ice and allow frozen reagents to thaw (see Note 4): DEPC-treated water (1), ammonium acetate (11), magnesium acetate (4), amino acid mixture (12), cell-free reaction supplement (19), creatine phosphate (13), creatine kinase (14), and T7 RNA polymerase (20).
2. Thaw DNA template solution (Subheading 3.2.1) allowing for in excess of 6 μg of DNA per 1 mL reaction.
3. Thaw an aliquot of S30 *E. coli* extract (Subheading 3.1) which constitutes 25% v/v of a reaction mixture.
4. A standard 1 mL cell-free reaction mixture is prepared by combining the following (final concentration shown in parenthesis): 311 μL DEPC-treated water; 20 μL ammonium acetate stock (28 mM); 11 μL magnesium acetate stock (15.4 mM); 40 μL amino acid mixture (1 mM each amino acid); 250 μL cell-free reaction supplement (25% v/v); 80 μL creatine phosphate stock (80 mM); 25 μL creatine kinase stock (250 μg/mL); 7 μL T7 RNA polymerase stock (77 μg/mL); 6 μL DNA template solution (6 μg/mL), and 250 μL S30 *E. coli* extract (25% v/v).
5. Gently mix the cell-free reaction mixture by inverting the tube several times and incubate at 37°C for 3 h (see Note 5).
6. Small aliquots should be taken at regular intervals during incubation for the analysis of expression by SDS–PAGE/western blot techniques (see Note 6).

This standard cell-free reaction recipe should be optimised for factors influencing transcription, including the concentration of RNA polymerase, DNA concentration, and the concentration of magnesium for expression of a given construct. Yield and solubility of the expressed protein can be strongly influenced by the temperature. 37°C is the standard reaction temperature. Lowering reaction temperature by as little as 2°C can significantly enhance protein solubility without significant impact on total yield.

Radiolabels, such as ^{35}S-methionine, can be added to cell-free reactions to rapidly determine optimum expression conditions from very small reaction volumes (see Fig. 1), where antibodies are not available. The lack of background expression of endogenous *E. coli* proteins means that autoradiography from dot blots can rapidly determine relative expression levels without the need for separation by techniques, such as SDS–PAGE.

Once optimal reaction conditions are identified, mg/mL level expression can be achieved by extending the life of the cell-free

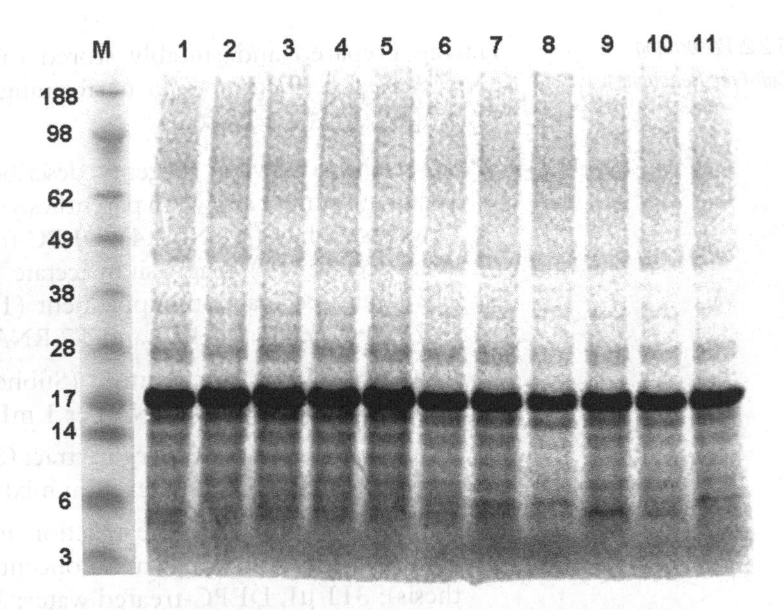

Fig. 1. Soluble fractions from cell-free reactions supplemented with ^{35}S labelled methionine expressing a Prion protein domain (PrP 91–231). Samples were separated by SDS–PAGE and visualised by autoradiography on a PhosphorImager. Each lane represents the relative levels of soluble protein expressed in the presence of critical micellular concentrations of detergents: (1) No detergent reference; (2) C12E9; (3) C12E8; (4) *n*-Dodecyl-β-D-maltoside; (5) Sucrose monolaurate; (6) CYMAL-6; (7) TRITON X-100; (8) CTAB; (9) Deoxy BigChap; (10) *n*-Decyl-β-D-maltoside; and (11) LDAO.

reaction by performing the reaction in *dialysis mode*. Here, the reaction is dialysed (10–15 kDa molecular weight cut off membrane) against 10 volumes of a buffer of the same composition as the reaction mixture but lacking *E. coli* extract, RNA polymerase, creatine kinase, and DNA template. In this way, in vitro expression can be sustained for roughly 8 h (37°C) or 24 h (30°C) of incubation (see Note 7).

4. Notes

1. It is our practice to prepare 6 L batches of *E. coli* culture. 2 L baffled conical flasks and 5-L baffled conical flasks have been used. 1 and 2 L of culture per flask, respectively, seem to provide suitable aeration. All flasks, beakers, and bottles (capped) are topped with a long aluminium sleeve or skirt so that the cap can be placed face down on a bench without the risk of contamination.
2. Every effort must be made to minimise RNase contamination at every stage in the preparation of cell-free extracts, when preparing solutions and when performing a cell-free reaction.

Once *E. coli* cells have been harvested, everything that the cells and extracts come into contact must have been treated to minimise RNase contamination to the greatest extent possible. When preparing stock solutions, it is best to use brand new laboratory bottles and high-quality plastic ware made from virgin materials. Clean glassware, magnetic stirrers, and utensils should be baked out in aluminium foil for several hours prior to their use. All laboratory ware must be washed with DEPC-treated water and autoclaved. *Stock solutions* should be made with DEPC-treated milli-Q water with 0.003% (w/v) sodium azide and sterile filtered (0.22 μm) when reagents cannot be autoclaved. Sterilisation filters should themselves be washed with DEPC treated water to remove glycerol and other residues. This does not include cell culture media, which should be made with untreated distilled water. Where possible, whole stock solutions should be prepared and DEPC treated.

3. Reagents for a cell-free reaction are thawed on ice and should be combined in the order described to prevent salting-out or denaturation of enzymes, such as creatine kinase or RNA polymerase. S30 extract is added just prior to initiating transcription/translation by heating to the reaction temperature (typically 30–37°C). Reactions are made to volume with DEPC-treated water, but this is added first and so the volume required must be pre-calculated to compensate for variation in the volumes of other reagents used, such as DNA, magnesium acetate, RNA polymerase solutions that need to be optimised.

4. Use a large volume of ice in an insulated chest to minimise wetting of the ice bed that can promote contamination of the reagents with proteases, RNases, etc.

5. Reactions can be performed in Eppendorf tubes or similar RNase-free/protease-free sterile laboratory tubes. These tubes should be sealed and wrapped with paraffin film (to avoid contamination) if incubated in a water bath fitted with a shaker for gentle agitation during reaction. Alternatively, short-lived batch reactions can be performed in sealed 96-well plates heated in a standard shaking incubator. One of the best configurations for batch mode optimisation is to perform reactions in polymerase chain reaction (PCR) tubes with temperature controlled in a PCR thermocycler.

6. The cell-free reaction mixture contains polyethylene 8000 which causes significant smearing of SDS–PAGE and should be removed prior to electrophoresis. Our practice is to vigorously mix (vortex) 1 volume of reaction mixture with 2 volumes of acetone to precipitate proteins. The mixture is subjected to centrifugation (13,000×*g*, 15 min), the supernatant

is then decanted off and the pellet is air dried and resuspended in SDS–PAGE loading buffer.

7. To achieve high-level expression in an extended (dialysis mode) reaction, RNase inhibitors are generally required. Inclusion of RNase inhibitors can add significantly to the expense of cell-free expression, and we find it cost-effective to exclude such inhibitors from initial batch mode optimisation experiments.

References

1. Miles, L. A. (2005) "Robust and cost-effective cell-free expression of biopharmaceuticals: Escherichia coli and Wheat Embryo" in Modern Biopharmaceuticals (Knäblein, J., Ed.), Vol. 3, pp. 1063–82, WILEY-VCH Verlag GmbH&Co. KGaA, Weinheim.
2. Grandi, G. (Ed.) (2007) In vitro transcription and translation protocols, Methods Mol Biol. 375, Humana Press, Totowa.
3. Kigawa, T., Yabuki, T., Yoshida, Y., Tsutsui, M., Ito, Y., Shibata, T., and Yokoyama, S. (1999) "Cell-free production and stable-isotope labeling of milligram quantities of proteins". FEBS Lett **442**, 15–9.
4. Kim, D. M., and Choi, C. Y. (1996) "A semi-continuous prokaryotic coupled ranscription/translation system using a dialysis membrane". Biotechnol Prog **12**, 645–9.
5. Kim, D. M., and Swartz, J. R. (2000) "Prolonging cell-free protein synthesis by selective reagent additions". Biotechnol Prog **16**, 385–90.
6. Spirin, A. S., Baranov, V. I., Ryabova, L. A., Ovodov, S. Y., and Alakhov, Y. B. (1988) "A continuous cell-free translation system capable of producing polypeptides in high yield". Science **242**, 1162–4.
7. Yokoyama, S., Matsuo, Y., Hirota, H., Kigawa, T., Shirouzu, M., Kuroda, Y., Kurumizaka, H., Kawaguchi, S., Ito, Y., Shibata, T., Kainosho, M., Nishimura, Y., Inoue, Y., and Kuramitsu, S. (2000) "Structural genomics projects in Japan". Prog Biophys Mol Biol **73**, 363–76.

Chapter 3

Synthesis of Peptide Sequences Derived from Fibril-Forming Proteins

Denis B. Scanlon and John A. Karas

Abstract

The pathogenesis of a large number of diseases, including Alzheimer's Disease, Parkinson's Disease, and Creutzfeldt–Jakob Disease (CJD), is associated with protein aggregation and the formation of amyloid, fibrillar deposits. Peptide fragments of amyloid-forming proteins have been found to form fibrils in their own right and have become important tools for unlocking the mechanism of amyloid fibril formation and the pathogenesis of amyloid diseases. The synthesis and purification of peptide sequences derived from amyloid fibril-forming proteins can be extremely challenging. The synthesis may not proceed well, generating a very low quality crude product which can be difficult to purify. Even clean crude peptides can be difficult to purify, as they are often insoluble or form fibrils rapidly in solution. This chapter presents methods to recognise and to overcome the difficulties associated with the synthesis, and purification of fibril-forming peptides, illustrating the points with three synthetic examples.

Key words: Solid-phase peptide synthesis, Fibril-forming peptide, Microwave peptide synthesis, Peptide HPLC purification, ApoCII [56–76], Aβ[1–42], Prion protein

1. Introduction

Solid-phase peptide synthesis (SPPS) methodology was first published in 1963 by Bruce Merrifield (1), wherein he described the synthesis of a tetrapeptide using a Boc/Bzl protecting group strategy on an insoluble polymer support. The technique was steadily developed during the 1970s and 1980s through automation of the peptide synthesis cycle and improvements to solid-phase Boc/Bzl chemistry (2, 3), together with the introduction of the solid-phase Fmoc/tBu protecting group strategy (4, 5). The introduction of new protected amino acid derivatives and activation reagents (6, 7) resulted in faster cycle times and higher quality crude products. These improvements enabled the synthesis

Andrew F. Hill et al. (eds.), *Protein Folding, Misfolding, and Disease: Methods and Protocols*, Methods in Molecular Biology, vol. 752, DOI 10.1007/978-1-60327-223-0_3,

of increasingly long and difficult sequences, such as the Boc/Bzl synthesis of HIV protease by Clark-Lewis et al. in 1986 (8) and the Fmoc/tBu synthesis of TGF-α by Scanlon et al. in 1987 (9).

The advancements in chemistry enabled the routine synthesis of polypeptides up to 50 amino acids in length, but the success of a step-wise synthesis of small proteins over 100 residues is sequence-dependent and therefore not universally achievable. To overcome this problem, Dawson et al. (10) developed a convergent synthetic approach (native chemical ligation), whereby a peptide fragment is synthesised with a C-terminal thioester residue which can then be condensed with a second peptide with an N-terminal cysteine residue. The cysteine thiol displacement of the thioester, followed by an S to N acyl shift to form an amide bond, results in the facile condensation of the two peptides. This technology has been used in conjunction with both Boc/Bzl and Fmoc/tBu chemistry (11), and has extended the scope of peptide synthesis to the extent that it is now a viable method for the chemical assembly of small proteins (see ref. 12 for a review.).

Despite these advances, the synthesis of fibril-forming peptides, such as the Aβ[1–42] peptide associated with Alzheimer's disease, is still problematic and presents a unique challenge for the peptide chemist. These sequences do not assemble well on the solid phase due to the formation of β-sheets and aggregation of the growing peptide chain on the resin. This results in either incomplete *N*-α Fmoc deprotection and/or incomplete coupling of the activated protected amino acid with the resin bound *N*-α amine (13). Methods reported to overcome these problems include the use of pseudo-prolines at critical points in the synthesis (14), or the incorporation of amide-backbone protecting groups, such as 2-hydroxy-4-methoxybenzyl (Hmb) or 2,4-dimethoxybenzyl (Dmb) at critical points in the synthesis to hinder aggregation on the resin (15). These novel chemical techniques result in significantly higher quality crude peptide products.

However, even though the sequence may assemble well on a solid phase resin, fibril-forming peptides are generally very poorly soluble in the aqueous buffers required for purification. Moreover, peptides have a high propensity to rapidly form fibrils, sometimes in as little as a few minutes, which makes handling extremely difficult (16). A reported solution to this problem is the introduction of a Ser(Xaa) or Thr(Xaa) isoacyl dipeptide at a critical point in the peptide sequence, which, on cleavage, generates a depsi-peptide that can hinder fibril formation. The native molecule may then be generated in aqueous buffer after purification via a pH dependent O to N acyl shift. This technique has been successfully applied to a synthesis of Aβ[1–42], where an Fmoc-Ser(Gly)-OH depsi-dipeptide has been inserted at position 25 and 26 (17, 18).

This chapter describes general methods to synthesise, cleave, isolate, and purify fibril-forming peptides. It is impossible to detail every possible synthetic method, purification technique or piece of instrumentation available for peptide synthesis, but care has been taken to cover a broad range of successful techniques, from the choice of amino acids and reaction conditions during synthesis, to buffer and column choice for purification. The use of microwave-assisted SPPS is emphasised and its efficacy discussed, as is the handling protocols of both the crude and pure peptides. The synthesis and purification of fibril-forming peptides are illustrated by three examples, namely, ApoCII [56–76], Aβ[1–42, $L^{34}(^{13}C)$, $V^{36}(^{15}N)$], and PrP[23–111]. The various strategies employed in order to optimise the yield and purity of these challenging peptides are discussed in detail.

2. Materials

2.1. Synthesis Reagents

1. *Amino acids*: Fmoc-l-amino acids and coupling reagents O-(6-chlorobenzotriazol-1-yl)-*N*,*N*,*N′*,*N′*-tetramethyluronium-hexafluorophosphate (HCTU) and *N*-hydroxybenzotriazole (HOBt) (GL Biochem; Shanghai, China).
2. *Resins*: Fmoc-l-AA-PEG resins and Fmoc-Pal-PEG-PS resin (Applied Biosystems; Melbourne, Australia).
3. *Solvents*: dimethylformamide (DMF), dichloromethane (DCM), acetonitrile, and diethyl ether (Ajax Pty Ltd; Melbourne, Australia).
4. *Reagents*: Trifluoroacetic acid (TFA), 3,6-dioxa-1, 8-octanedithiol (DODT), triisopropylsilane (TIPS), piperidine, diisopropylethyl amine (DIPEA), *N*-methylpyrrolidone (NMP) (Sigma Aldrich; Sydney, Australia).

2.2. Synthesis Instrumentation

1. *Automated Peptide Synthesiser*. CEM Liberty Microwave Peptide Synthesiser.
2. *Cleavage reactions*. Eppendorf centrifuge 5702.
3. *Lyophilisation*. Virtis benchtop SLC freeze drying system.

2.3. Chromatography Reagents

1. Acetonitrile.
2. Isopropanol.
3. TFA.
4. Formic acid.
5. Acetic acid (Glacial).

2.4. Chromatography Instrumentation

2.4.1. Analytical System

1. Agilent 1100 G1311A Quaternary Pumping System.
2. Agilent 1100 G1313A Autosampler.
3. Agilent 1100 G1315B Diode-array detector.
4. Agilent 1100 G1316A Column Heater.

2.4.2. Preparative System

1. Agilent 1200 G1311A Quaternary Pumping System.
2. Agilent 1200 G1365B Multi-Wavelength Detector.
3. Agilent 1200 G1364C Fraction Collector.
4. Rheodyne Valve 7725i with 5 ml sample loop.

2.5. Columns

2.5.1. Analytical System

1. Agilent Zorbax 5 μm C_{18} , 4.6×250 mm analytical column.
2. Vydac 10 μm C_4, 4.6×250 mm analytical column.

2.5.2. Preparative System

1. Agilent Zorbax 5 μm C_{18}, 10×250 mm semi-preparative column.
2. Vydac 10 μm C_4, 25×250 mm preparative column.

2.6. HPLC Buffers

2.6.1. TFA Buffer pH2: General

Buffer A: 0.1% aqueous TFA.

Buffer B: 90% acetonitrile/water (containing 0.1% TFA).

2.6.2. TFA Buffer pH2: Aβ Peptides/Hydrophobic Peptides

Buffer A: 0.1% aqueous TFA.

Buffer B: 50% acetonitrile/50% isopropanol (containing 0.1% TFA).

2.7. Mass Spectrometry Instrumentation

Agilent 6220 accurate-mass TOF LC/MS (ESI-TOF MS).

3. Methods

A flow chart outlining peptide production, which describes each process from the solid-phase synthesis through to the final QC of the purified material, is shown in Fig. 1. During the process, decisions need to be made regarding synthesis protocols, purification conditions and handling of the peptide during the process to prevent fibril formation. These issues are discussed in turn below.

3.1. Synthesis on CEM Liberty Synthesiser

The chemist should be familiar with Fmoc SPPS protocols before commencing a difficult synthesis on an automated instrument. In order to achieve optimum results, low-loading pegylated resins, which solvate well in DMF, were routinely used in a microwave reactor.

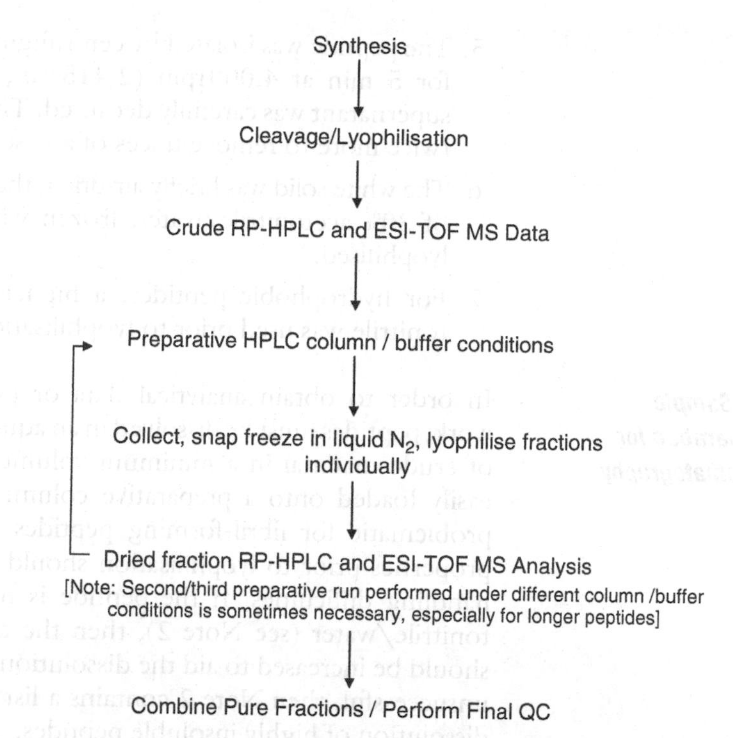

Fig. 1. Stepwise process of fibril-peptide synthesis and purification.

1. The CEM Liberty instrument uses a 40 min cycle which includes 4 min 25% piperidine/DMF Fmoc deprotection at 75°C, followed by an Fmoc protected amino acid coupling time of 5 min at 75°C.
2. All Fmoc-L-Arg(Pbf)-OH residues were double coupled, as were Fmoc-L-His(Trt)-OH and Fmoc-L-Cys(Trt)-OH, but at 50°C to minimise racemisation.
3. For longer peptides, the coupling time was increased to 15 min at 75°C after 30 cycles.

3.2. Cleavage of the Resin-Bound Peptide and Lyophilisation

1. Cleavage of resin-bound peptides was performed in 50 ml Falcon tubes with a ratio of 10 mL of cleavage reagent per 1 g of dry resin.
2. The cleavage cocktail consisted of TFA:TIPS:water:DODT (92.5:2.5:2.5:5, 10 mL), and the reaction mixture was gently agitated for 2–4 h.
3. The solution was filtered (see Note 1), and the filtrate was concentrated to approximately 1 ml with a stream of dry nitrogen.
4. Cold diethyl ether was then added and the crude material precipitated as a white solid.

5. The peptide was isolated by centrifuging the cleavage solution for 5 min at 4,000 rpm (2,415 × *g*), after which the ether supernatant was carefully decanted. This process was repeated twice more to remove traces of any scavengers.
6. The white solid was briefly air-dried, then dissolved in a mixture of 30% acetonitrile/water, frozen with liquid nitrogen and lyophilised.
7. For hydrophobic peptides, a higher concentration of acetonitrile was used prior to lyophilisation (see Note 2a).

3.3. Sample Preparation for Chromatography

In order to obtain analytical data or perform any preparative work, peptides must be dissolved in an aqueous buffer. Dissolution of crude material in a minimum volume enables samples to be easily loaded onto a preparative column; however, this can be problematic for fibril-forming peptides. The peptide solubility properties prior to lyophilisation should provide a clue as to its handling difficulties. If the peptide is not soluble in 30% acetonitrile/water (see Note 2), then the acetonitrile composition should be increased to aid the dissolution of the peptide. If this is unsuccessful, then Note 2 contains a list of alternate options for dissolution of highly insoluble peptides.

3.4. Analytical Data

1. The first critical step in the purification process is to obtain an RP-HPLC chromatogram and mass spectrum of the crude material.
2. Typically, an analytical C_{18} reversed-phase column (e.g. Agilent Zorbax 5 μm C_{18} 4.6 × 250 mm analytical column) is employed, using a flow rate of 1 mL/min., with peak detection at 220 nm and a linear gradient from 0 to 60% Buffer B over 30 min (~2% per minute) in the TFA buffer system (pH 2; see Note 3).
3. Peptides will generally elute between 10 and 25 min under these conditions, however, if they elute later or run as an extremely broad peak, then heating the column at 60°C will shorten the retention time and sharpen the peak. This allows peptide purities to be more accurately measured.
4. Software packages, such as Peptide Companion (see Note 4), can calculate the monoisotopic and average molecular weights.
5. An ESI-TOF mass spectrometer (see Subheading 2.7) can accurately measure the monoisotopic mass of the crude peptide, thus confirming unambiguously the integrity of the peptide sequence.

3.5. Preparative Chromatography

3.5.1. System Set-Up

1. The preparative conditions are a scale up of those used to generate the analytical data.
2. RP-HPLC columns, such as an Agilent Zorbax 5 μm C_{18}, 10 × 250 mm semi-prep column, have the capacity to purify peptides in 20–50 mg batches.

3. Gradients between 0.5 and 1% per minute Buffer B are typically employed in order to optimise separation. Peaks are detected between 224 and 230 nm, depending on sample amount (see Note 5).
4. The sample is introduced onto the column via a rheodyne valve fitted with a 5 mL sample loop (see Note 6).
5. The fraction collector is programmed to begin collecting the eluent when it detects an absorbance of a preset threshold in the detector (see Note 7).
6. Collected fractions are frozen in 15 mL falcon tubes and individually analysed and lyophilised.

3.5.2. Fraction Analysis

1. Small aliquots from the fractions are analysed by the ESI-TOF mass spectrometry.
2. Once the target has been identified, the surrounding fractions can then be RP-HPLC analysed, and the pure fractions (usually >90–95% purity) are combined.
3. It is good practice to pool and re-lyophilise the clean fractions for a consistent product, which should be reanalysed via RP-HPLC and ESI-TOF mass spectrometry.
4. The pure peptide can then be stored dry at 4°C or −20°C, depending on the stability of the peptide.

3.6. Examples

Three examples have been selected to illustrate the challenges that can be encountered during the synthesis, purification, and analysis of fibril-forming peptides and also to highlight various methods which can be used to overcome these difficulties.

3.6.1. ApoCII [56–76]

ApoCII is a small apolipoprotein that has been associated with atherosclerosis in blood. It has been reported that, together with other amyloidic apolipoproteins, it is responsible for the deposition of plaques in arteries causing atherosclerotic lesions which may be implicated in heart disease pathogenesis (19). The study of this process has identified a 21 amino acid peptide subsequence of ApoCII, namely, ApoCII[56–76], which forms fibrils and is thought to be responsible for the aggregation properties of ApoCII itself (20).

ApoCII[56–76] is relatively easy to synthesise, but is quite insoluble and readily forms fibrils in aqueous buffers, making it extremely difficult to purify. Therefore, in order to simplify the purification, the methionine residue at position 60 was synthesised as the methionine-sulfoxide (Met[O]). This increased the solubility of the peptide and diminished its aggregation properties. It can be seen in Fig. 2a that the RP-HPLC trace of the Met[O] version of the ApoCII[56–76, Met ^{60} [O]] peptide contains a single major product, which runs as a sharp peak. The mass spectrum of the crude peptide in Fig. 2b is in accord with the

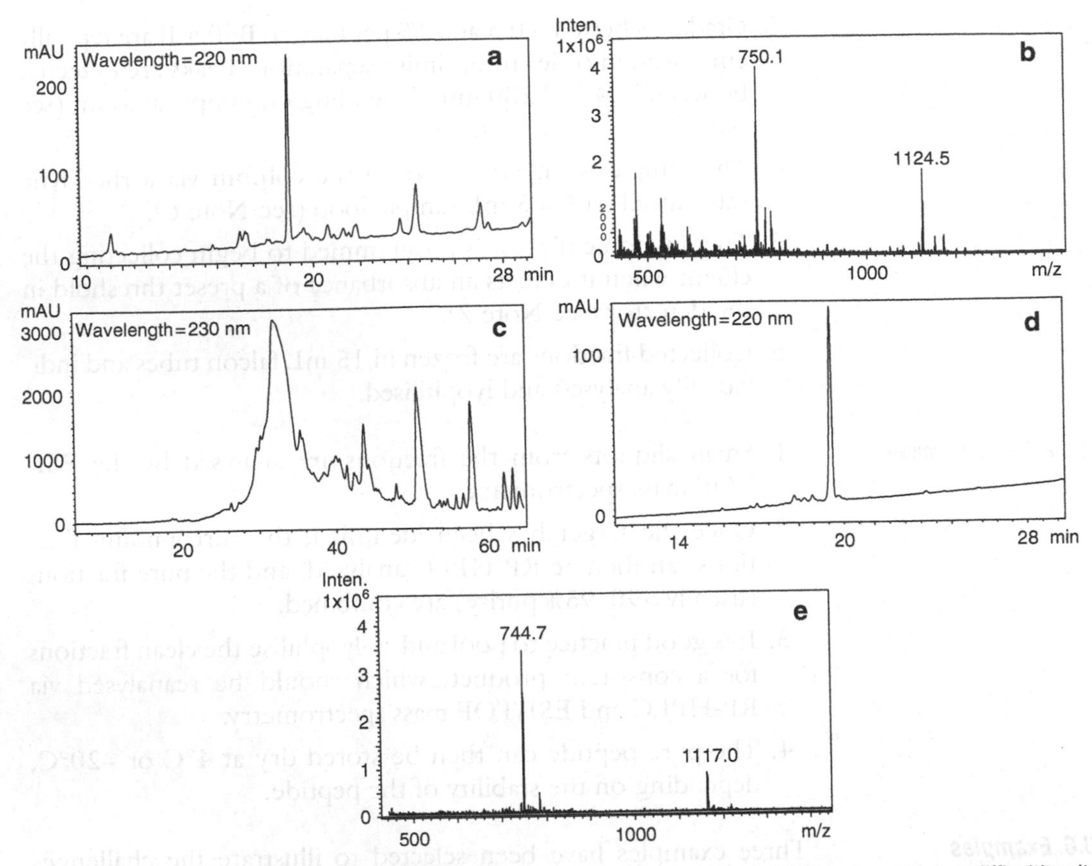

Fig. 2. ApoCII[56–76] was synthesised with an Met[O] derivative to overcome solubility and purification difficulties. It was reduced at the end of the purification process to generate the natural sequence (**a**) ApoCII[56–76, Met[60][O]] crude analytical RP-HPLC trace (**b**) ApoCII[56–76, Met[60][O]] crude mass spectrum showing $[M+2H]^{2+}$ and $[M+3H]^{3+}$ ions, MW: 2,247.3 Da (**c**) ApoCII[56–76, Met[60][O]] preparative RP-HPLC chromatogram and mass spectrum (**d**) Pure ApoCII[56–76] RP-HPLC trace after the treatment with TMSBr to reduce the Met[O] to Met (**e**) Pure ApoCII[56–76] ESI-TOF MS data showing $[M+2H]^{2+}$ and $[M+3H]^{3+}$ ions, MW: 2,231.1 Da.

theoretical molecular weight of 2,247.2. The enhanced aqueous solubility and monomeric integrity ensures that purification is straightforward, using the protocols described above (see Fig. 2c). The isolated, purified Met[O] peptide was easily and quantitatively converted to the native reduced, form by the treatment with trimethylsilyl bromide in TFA (see Note 8) without further purification. Quality control on the final lyophilised product verified the peptide's high purity by RP-HPLC (Fig. 2d), and confirmed its identity via ESI-TOF MS data (MW: 2,231.2 – Fig. 2e).

3.6.2. Amyloid-β Peptide (Aβ[1–42, Leu[34] (^{13}C), V[36] (^{15}N)])

The Aβ[1–42] peptide has been identified as a key element in the pathogenicity of Alzheimer's disease due to its ability to aggregate into fibril plaques in the brain and its neurotoxicity (21). It has been well documented in the literature as a difficult sequence to synthesise by standard peptide synthesis methodologies (13, 14) and there are reports of novel chemical methods to overcome the synthesis difficulties (16–18). However, difficult to assemble peptides, such as Aβ[1–42] can be synthesised in high purity by microwave peptide synthesis technology.

The peptide Aβ[1–42, Leu^{34} (^{13}C), V^{36} (^{15}N)], synthesised for solid-state NMR experiments, was assembled on the CEM Liberty Microwave Peptide Synthesiser under the conditions described in Subheading 3.

1. Fmoc-l-Leu^{34}(^{13}C)-OH and Fmoc-l-Val^{36}(^{15}N)-OH were coupled manually with a twofold excess to minimise the usage of these costly amino acid derivatives. All other amino acids were incorporated using automated cycles.
2. The quality of the crude peptide is illustrated in the ESI-TOF MS data (Fig. 3a), which clearly shows the $[M+3H]^{3+}$, $[M+4H]^{4+}$, and $[M+5H]^{5+}$ charge states associated with the parent mass of 4,513.3 Da.
3. Due to the difficulty in handling amyloid beta peptides in solution, small-scale repetitive semi-preparative purifications were performed (see Fig. 3b) in order to keep concentrations low so that aggregation was minimised.

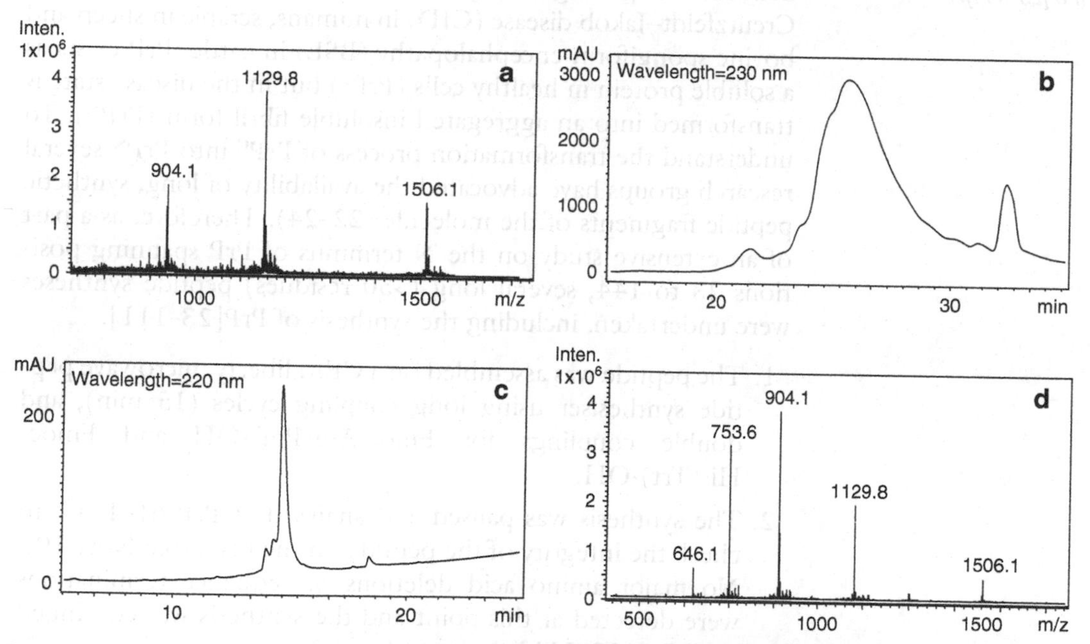

Fig. 3. The peptide Aβ[1–42, Leu^{34} (^{13}C), V^{36} (^{15}N)] was synthesised for solid-state NMR experiments and exhibited a molecular weight two mass units higher than the native peptide (4,511.28 Da) (**a**) Crude mass spectrum of Aβ[1–42, Leu^{34} (^{13}C), V^{36} (^{15}N)] showing $[M+3H]^{3+}$ to $[M+5H]^{5+}$ ions: MW = 4,513.3 Da (**b**) Preparative RP-HPLC trace on a Vydac C_4 column run at 60°C. The profile was manually fractionated into 1 min fractions which were snap frozen in liquid nitrogen and lyophilised. (**c**) Pure analytical chromatogram using an analytical Vydac C_4 reversed-phase HPLC column at 60°C. (**d**) Pure analytical ESI-TOF MS data showing $[M+3H]^{3+}$ to $[M+7H]^{7+}$ showing the correct molecular weight (4,513.3 Da).

4. A Vydac C_4 preparative column was preferred because Aβ[1–42, Leu^{34} (^{13}C), V^{36} (^{15}N)] does not elute off the more hydrophobic C_{18} column.
5. Buffer B comprised 50% acetonitrile/50% isopropanol (see Subheading 2.6.2), and the chromatography was performed with a column temperature of 60°C. These conditions enhance the sharpness of the peak and decrease the retention time.
6. Fractions were hand collected (1 min fractions) and immediately frozen in liquid nitrogen and lyophilised, to avoid fibril formation (see Note 9).
7. Four individual runs of approximately 30 mg crude peptide each were performed in this manner and the fractions, once dry, were analysed by ESI-TOF mass spectrometry for purity.
8. Fractions with the "cleanest" mass spectra were combined, and reanalysed via RP-HPLC on a C_4 column at 60°C (Fig. 3c), and ESI-TOF MS (Fig. 3d). As expected, a molecular weight of 4,513.3 was measured, which is 2 Da higher than the native peptide, due to the ^{13}C and ^{15}N amino acid substitutions.

3.6.3. Prion Protein (PrP[23–111])

An abnormal isoform of the prion protein (PrP) has been identified as the pathogenic species in the neurological diseases Creutzfeldt–Jakob disease (CJD) in humans, scrapie in sheep and bovine spongiform encephalopathy (BSE) in cattle. PrP exists as a soluble protein in healthy cells (PrP^{C}) but in the disease state is transformed into an aggregated insoluble fibril form (PrP^{Sc}). To understand the transformation process of PrP^{C} into PrP^{Sc} several research groups have advocated the availability of long, synthetic peptide fragments of the molecule (22–24). Therefore, as a part of an extensive study on the N-terminus of PrP spanning positions 23 to 144, several long (>50 residues) peptide syntheses were undertaken, including the synthesis of PrP[23–111].

1. The peptide was assembled on a CEM liberty microwave peptide synthesiser using long coupling cycles (15 min), and double couplings for Fmoc-Arg(Pbf)-OH and Fmoc-His(Trt)-OH.
2. The synthesis was paused and analysed at PrP[61–111], to check the integrity of the peptide on the resin (see Note 10). No major amino acid deletions or sequence terminations were detected at this point and the synthesis was continued until PrP[23–111] was complete.
3. After cleavage, the crude material was lyophilised three times to remove all traces of the Boc protecting group, originating from the five tryptophan residues present in the peptide (see Note 11).

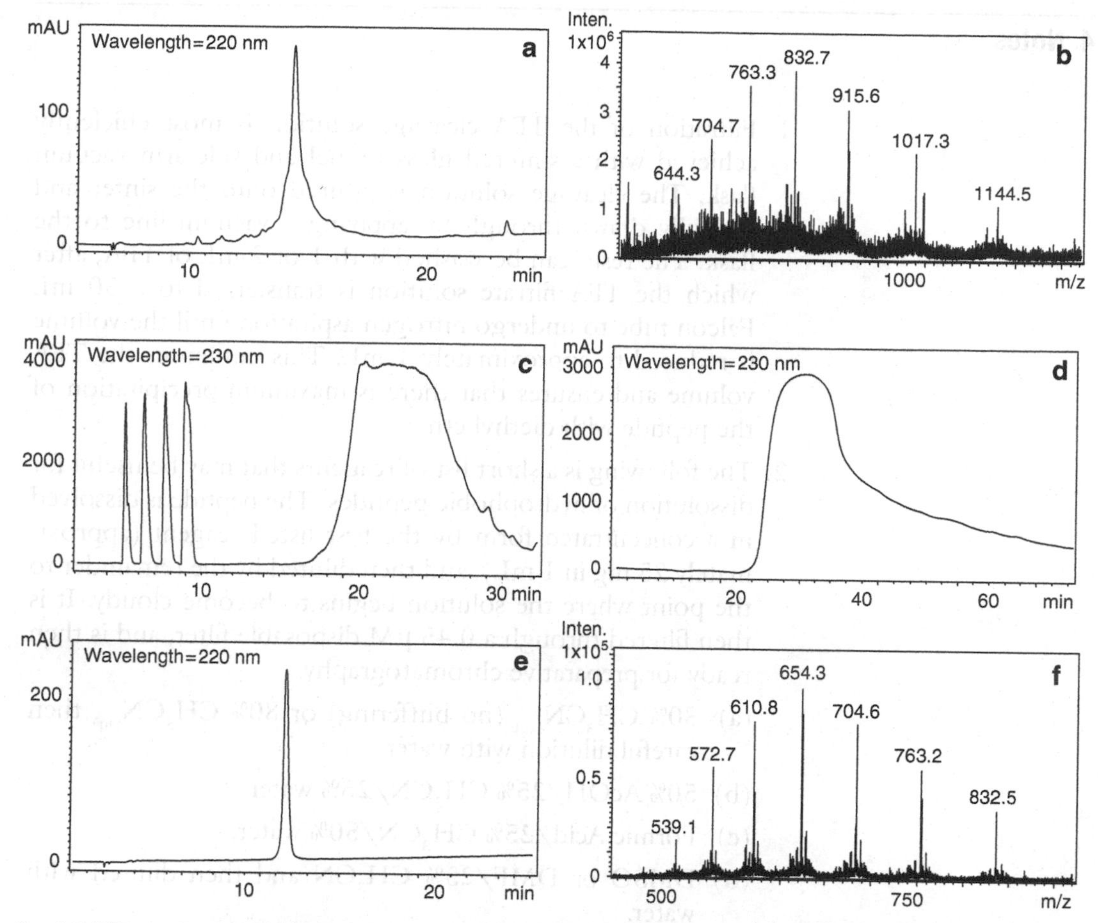

Fig. 4. PrP[23–111] is the N-terminal segment of the fibril-forming protein Prion Protein. (**a**) PrP[23–111] crude RP-HPLC chromatogram, (**b**) Crude ESI-TOF MS data showing $[M+8H]^{8+}$ to $[M+14H]^{14+}$ ions, MW: 9,141.4 Da (**c**) PrP[23–111] first preparative RP-HPLC on a Vydac C_4 preparative column (**d**) Second preparative RP-HPLC chromatogram on a Zorbax C_{18} semi-preparative column. (**e**) PrP[23–111] pure RP-HPLC chromatogram. (**f**) Pure ESI-TOF MS data showing $[M+11H]^{11+}$ to $[M+17H]^{17+}$ ions, MW: 9,141.4 Da.

4. The crude RP-HPLC trace and the crude ESI-TOF MS data are shown in Fig. 4a, b. The mass spectral data confirms that PrP[23–111] with a molecular weight of 9,141.4 is the major product.
5. The peptide was initially purified to >85% purity by RP-HPLC preparative runs on a Vydac C_4 column, followed by a final purification using a Zorbax C_{18} semi-preparative column.
6. The pure RP-HPLC chromatogram in Fig. 4e shows the PrP[23–111] as a single homogeneous peak. Figure 3.4f shows the pure ESI-TOF MS data, where the $[M+11H]^{11+}$ to $[M+17H]^{17+}$ ions are apparent, corresponding to the expected molecular weight of 9,141.4.

4. Notes

1. Filtration of the TFA cleavage solution is most efficiently achieved with a sintered glass funnel and side-arm vacuum flask. The cleavage solution is poured onto the sinter and carefully drawn through by applying a vacuum line to the flask. The resin can be washed with 1 or 2 mL of TFA, after which the TFA filtrate solution is transferred to a 50 mL Falcon tube to undergo nitrogen aspiration until the volume is reduced to approximately 1 mL. This minimises the TFA volume and ensures that there is maximum precipitation of the peptide with diethyl ether.
2. The following is a short list of reagents that may be useful for dissolution of hydrophobic peptides. The peptide is dissolved in a concentrated form by the first listed reagent (approximately 25 mg in 1 mL), and then diluted by the remainder to the point where the solution begins to become cloudy. It is then filtered through a 0.45 μM disposable filter, and is then ready for preparative chromatography.
   (a) 30% $CH_3CN_{(aq)}$ (no buffering) or 80% $CH_3CN_{(aq)}$ then careful dilution with water.
   (b) 50% AcOH/25% CH_3CN/25% water.
   (c) Formic Acid/25% CH_3CN/50% water.
   (d) DMSO or DMF/25% CH_3CN and then diluted with water.
   (e) 8 M Urea or 7 M Guanidine. HCl can be useful if the above methods fail.
3. In most cases, TFA buffers are used because of their resolving power, and because it is volatile during lyophilisation. To prepare a 0.1%TFA/water buffer measure 990 mL of water and add 10 ml of a 10% TFA/water solution. This is preferable to using a micropipette to measure small volumes of neat TFA due to accuracy and handling concerns. A 500 ml stock solution of 10% TFA/water can be prepared and stored ready for use.
4. Peptide Companion from CSPS Pharmaceuticals is a useful software package which calculates various properties of peptides, such as molecular weight and can provide hydrophobicity/hydrophilicity plots and RP-HPLC retention time predictions.
5. It is best practice to tune the signal out from 220 nm for detection of the amide bond, between 224 and 230 nm, otherwise the detector becomes overloaded and the data is of a poor quality. If tryptophan or tyrosine is present in the peptide, then an absorbance of 275–285 nm can be used as the detection wavelength. The preparative trace will be broad,

most likely off-scale, and show at best a slight resemblance to the analytical trace. Fraction analysis confirms that resolution was achieved, but it may not be obvious from the preparative trace.

6. The sample solution *must* be filtered preferably through a 0.45 μM disposable filter to protect the sinter on the top of the HPLC column. The acetonitrile concentration of the peptide sample solution must be approximately 5% below the concentration at which the peptide is eluted from the column in its analytical run. As fibril-forming peptides are poorly soluble, the sample volume is generally greater than the size of the loop attached to the rheodyne valve. This problem is overcome by inserting a 5 minute isocratic step at the start of the prep run before the commencement of the gradient and performing multiple injections of the sample solution. For example, if the peptide is dissolved in 15 mL and the loop size is 5 mL, then three injections of 5 mL with a 1–1.5 min lag time between injections (at 5 mL per minute, 1 min is the time required to clear the sample from the loop) will successfully introduce all the sample on to the column.
7. Care must be taken when inputting the threshold value for fraction collector, as an unexpected low absorbance deflection during the preparative run may not trigger collection, and the product flows to waste.
8. Met[O] can be reduced to Met by treating the peptide with trimethylsilyl bromide/DODT/TFA (1:1:98) at 0°C for 15 min at a concentration of 20 mg/mL. The peptide is then precipitated with cold diethyl ether and isolated by centrifugation. The precipitation should be repeated three times to remove all reagents from the peptide.
9. When handling all Aβ[1–42] peptides, the time they remain in solution during purification must be kept to a minimum to prevent fibril formation. All fractions were, therefore, hand-collected and immediately frozen in liquid nitrogen and lyophilised.
10. The quality of the synthesis was checked by taking a small resin sample (10 mg) and cleaving the peptide as described in the Materials and Methods. The peptide was ether precipitated twice and analysed by the ESI-TOF mass spectrometry to check for the correct mass for PrP[61–111]. This technique has been routinely used in the production of other long polypeptides.
11. Tryptophan has a Boc protecting group on the indole ring sidechain and TFA deprotection cleaves this in a two stage process, whereby tBu is quickly cleaved, but the CO_2 remaining (appearing as +44 mass units in the mass spectrum) is slowly hydrolysed once the peptide is in aqueous buffer. It is

usually no longer present after lyophilisation, but when there are several tryptophan residues (as is the case for PrP[23–111]), then two or more lyophilisation treatments may be necessary for complete deprotection.

Acknowledgements

We thank Ms Keyla Perez for her advice with the Vydac C_4 preparative column technology and peptide purifications performed at 60°C.

References

1. Merrifield, R. B. (1963) Solid Phase Synthesis I. The Synthesis of a Tetrapeptide. *J. Amer. Chem. Soc.* **85**, 2149–2154.
2. Kent, S. B., Mitchell, A. R., Engelhard, M. and Merrifield, R. B.(1979) Mechanisms and prevention of trifluoroacetylation in solid-phase peptide synthesis. *Proc. Natl. Acad. Sci. U S A.* **76(5)**, 2180–2184.
3. Sarin, V. K., Kent, S. B., Tam, J. P. and Merrifield, R. B. (1981) Quantitative monitoring of solid-phase peptide synthesis by the ninhydrin reaction *Anal. Biochem.* **117(1)**, 147–157.
4. Brown, E., Sheppard, R. C. and Williams, B. J. (1983) Peptide Synthesis. Part 5. Solid-phase Synthesis of [15-Leucine] Little Gastrin. *J. Chem. Soc. Perkin Trans.* **I**, 1161–1167.
5. Sheppard, R. C. (1986) Modern methods of solid-phase peptide synthesis. *Science Tools* **33**, 9–16.
6. Schnolzer, M., Alewood, P., Jones, A., Alewood, D. and Kent, S. B. H. (1992) *In situ* neutralization in Boc-chemistry solid phase peptide synthesis. *Int. J. Peptide Protein Res.* **40**, 180–193.
7. Alberico, F. and Carpino, L. A. (1997) Coupling reagents and activation *Methods Enzymol.* **289**, 104–126.
8. Clark-Lewis, I., Aebersold, R., Ziltener, H., Schrader, J. W., Hood, L. E. and Kent, S. B. H. (1986) Automated Chemical Synthesis of a Protein Growth Factor for Hemopoietic Cells, Interleukin-3. *Science* **231**, 134–139.
9. Scanlon, D. B., Eefting, M. A., Lloyd, C. J., Burgess, A. W. and Simpson, R. J. (1987) Synthesis of Biologically Active Transforming Growth Factor-alpha by Fluorenylmethoxycarbonyl Solid Phase Peptide Chemistry. *J. Chem. Soc. Chem. Commun.* 516–518.
10. Dawson, P. E., Muir, T. W., Clark-Lewis, I. and Kent, S. B. (1994) Synthesis of proteins by native chemical ligation. *Science* **266**, 776–779.
11. Yamamoto, N., Tanabe, Y., Okamoto, R., Dawson, P.E. and Kajihara, Y. (2008) Chemical Synthesis of a Glycoprotein Having an Intact Human Complex-Type Sialyloligosaccharide under the Boc and Fmoc Synthetic Strategies *J. Am. Chem. Soc.* **130(2)**, 501–510.
12. Macmillan, D. (2006) Protein Synthesis: Evolving Strategies for Protein Synthesis Converge on Native Chemical Ligation Angew. *Chem. Int. Ed.* **45**, 7668–7672.
13. Tickler, A. K.,. Clippingdale, A. B. and Wade, J. D. (2004) Amyloid-β as a "Difficult Sequence" in Solid Phase Peptide Synthesis. *Protein & Peptide Letters* **11(4)**, 377–384.
14. Mutter, M., Nefzi, A., Sato, T., Sun, X., Wahl, F., and Wohr, T. (1995) Pseudo-prolines for accessing "inaccessible" peptides *Peptide Research* **8(3)**, 145–153.
15. Simmonds, R. G. (1996) Use of the Hmb backbone-protecting group in the synthesis of difficult sequences. *Int. J. Pept. Protein Res.* **47(1–2)**, 36–41.
16. Tickler, A. K., Barrow, C. J. and Wade, J. D. (2001) Improved Preparation of Amyloid-Peptides Using DBU as N-Fmoc Deprotection Reagent *J. Peptide Sci.* 7, 488–494.
17. Sohma, Y. and Kiso, Y. (2006) "Click peptides"-chemical biology-oriented synthesis of Alzheimer's disease-related amyloid beta peptide (abeta) analogues based on the "O-acyl isopeptide method". *Chembiochem.* **7(10)**, 1549–1557.
18. Taniguchi, A., Sohma, Y., Hirayama, Y., Mukai, H., Kimura, T., Hayashi, Y., Matsuzaki, K., Kiso, Y. (2009) "Click peptide": pH-triggered in situ production and aggregation of

monomer Abeta1-42. *Chembiochem.* **10(4)**, 710–715.

19. Howlett, G. J., and Moore, K. J. (2006) Untangling the role of amyloid in atherosclerosis *Current Opinion in Lipidology* **17**, 541–547.
20. Wilson, L. M., Mok, Y. F., Binger, K. J., Griffin, M. D., Mertens, H. D., Lin, F., Wade, J. D., Gooley, P. R. and Howlett, G. J. (2007) A structural core within apolipoprotein C-II amyloid fibrils identified using hydrogen exchange and proteolysis. *J. Mol. Biol.* **366(5)**, 1639–1651.
21. Van Nostrand , W.E., Davis-Salinas, J. and Saporito-Irwin, S. M. (1996) Amyloid beta-protein induces the cerebrovascular cellular pathology of Alzheimer's disease and related disorders *Ann. N Y. Acad. Sci.* 777, 297–302.
22. Ball, H. L. and Mascagni, P. (1996) Chemical protein synthesis and purification: a methodology. *Int. J. Peptide Protein Res.* **48**, 31–47.
23. Bonetto, V., Massignan, T., Chiesa, R., Morbin, M., Mazzoleni, G., Diomede, L., Angeretti, N., Colombo, L., Forloni, G., Tagliavini, F., and Salmona, M. (2002) Synthetic miniprion PrP106 *J. Biol. Chem.* **277**, 31327–31334.
24. Bahadi, R., Farrelly, P. V., Kenna, B. L., Kourie, J. I., Tagliavini, F., Forloni, G. and Salmona, M. (2003) PrP(82–146) homologous to a 7-kDa fragment in Channels formed with a mutant prion protein diseased brain of GSS patients *Am. J. Physiol. Cell. Physiol.* **285**, 862–872.

Chapter 4

Refolding Your Protein with a Little Help from REFOLD

Jennifer Phan, Nasrin Yamout, Jason Schmidberger, Stephen P. Bottomley, and Ashley M. Buckle

Abstract

The expression and harvesting of proteins from insoluble inclusion bodies by solubilization and refolding is a technique commonly used in the production of recombinant proteins. Despite the importance of refolding, publications in the literature are essentially ad hoc reports consisting of a dazzling array of experimental protocols and a diverse collection of buffer cocktails. For the protein scientists, using this information to refold their protein of interest presents enormous challenges. Here, we describe some of the practical considerations in refolding and present several standard protocols. Further, we describe how refolding procedures can be designed and modified using the information in the REFOLD database (http://refold.med.monash.edu.au), a freely available, open repository for protocols describing the refolding and purification of recombinant proteins.

Key words: Protein expression, Refolding, Renaturation, Inclusion body, Aggregate, Refold, Misfolding, Solubilization

1. Introduction

Biomedical and biotechnical research often involves the need to purify recombinant proteins in the simplest and most efficient manner possible while maximizing both the yield and quality of protein purified. The use of recombinant techniques and bacterial systems facilitates the expression of proteins on a large scale; however, a key limitation of such systems is often the insolubility of the target protein, which may be expressed largely as nonfunctional aggregates in inclusion bodies (1, 2). Despite the development of various growth conditions, bacterial strains, expression systems, and solubilizing fusion partners to increase and maximize protein solubility (1, 3–5), for some proteins these strategies still prove to be ineffective or highly inefficient. On the other

Andrew F. Hill et al. (eds.), *Protein Folding, Misfolding, and Disease: Methods and Protocols*,
Methods in Molecular Biology, vol. 752, DOI 10.1007/978-1-60327-223-0_4,

hand, the overexpression of insoluble proteins can be exploited by the fact that proteins produced in inclusion bodies are often very pure. As such, the solubilization and unfolding of aggregated proteins, followed by refolding and a simple, one-step purification, either sequentially or concurrently, in many cases proves to be the most direct and effective method of producing highly purified protein.

There is a plethora of documented and anecdotal data regarding the techniques of refolding proteins in vitro. The various procedures and methods involved have been extensively reviewed (6–11); however, until recently, there has been no logical process by which optimal conditions may be gleaned for proteins with specific characteristics. Thus, for a researcher working with a novel protein, finding the most suitable conditions for expression, solubilization, and refolding of proteins, a priori can be a relatively random process. However, a large proportion of the refolding literature has been recently deposited and structured into the REFOLD database (http://refold.med.monash.edu.au) (12–14). REFOLD is thus a valuable resource for researchers in developing new protocols for the purification of proteins. As of June 2009, REFOLD contains the details of more than 1,100 published protocols involving the overexpression, solubilization, and refolding of more than 750 recombinant proteins (belonging to more than 250 structural families and more than 320 organisms). Entries are also annotated with data relating to the properties of the protein, such as structural data, isoelectric point, oligomeric state, and the presence of disulphide bonds, as well as references and links to other knowledge databases.

REFOLD provides a detailed catalogue of successful refolding and purification methods for a wide range of proteins in a readily accessible and easy to read format. Annotation allows the relationships between protein characteristics and refolding protocols to be delineated. Because of the wealth and breadth of data, and the emergence of trends and patterns, the data in REFOLD are an excellent resource for the rational design of refolding protocols, as well as an ideal framework in which to discuss refolding methodologies in general.

2. Materials

Wash buffer: to remove remaining loosely bound proteins and cellular debris. This buffer may contain low concentrations (0.01–2%) of non-ionic detergents (e.g. Triton-X100, SDS) and/or low concentrations of denaturant (up to 2 M urea), and the final wash may be buffer only without detergent.

Solubilization buffer: to unfold aggregated protein. This buffer should contain a denaturing agent, most commonly, 4–6 M guanidine hydrochloride or 8 M urea. Alternative denaturants may include non-ionic detergents, such as *N*-lauroylsarcosine (0.3–5% (v/v)) (see Note 1).

Refolding buffer: to remove denaturant and promote folding, via buffer-exchange. Typically, a cocktail of pH buffer (Tris–HCl at pH range 7–9), salts (typically, 50–500 mM NaCl), redox agents (most commonly GSH/GSSG), and additives (see Table 1). Subheading 3.4 describes other buffer choices reported in REFOLD.

Table 1
Additives used in refolding buffers, indicating their popularity

Additive	No. REFOLD entries (%)	Concentrations used	Effect
α-cyclodextrin	0.08	4.8 mM	Stabilizer
ß-cyclodextrin	0.91	4.8–16.3 mM	Stabilizer
Ammonium sulphate	0.98	0.2 mM to 0.5 M	Salt
1% (v/v) Tween	0.08	–	Detergent
Brij 35	0.23	0.01–20%	Detergent
Calixarene	0.15	–	Calixarene
CHAPS	0.91	20 mM to 0.5 M	Detergent
Ethylene glycol	1.06	5–50%	Stabilizer
Glycerol	12.56	100 mM to 2.5 M	Stabilizer
Glycine	1.21	50 mM to 1 M	Glycine
L-Arginine	15.15	100 mM to 2.5 M	Stabilizer
Lauryl maltoside	0.08	500 mM	Detergent
Mannitol	0.15	0.10%	Stabilizer
N-substituted *N*-methylimidazolium cations	0.15	–	Stabilizer
Octylglucoside (OG)	0.08	20 mg/ml	Detergent
poly-*N*-isopropyl acrylamide (PNIPAAm)	0.15	–	Stabilizer
Polyethylene glycol (PEG)	0.60	0.2–0.5 M	Osmolyte
Sodium dioctyl sulfosuccinate (AOT)	0.08	400 mM	Detergent
Triton X-100	2.75	0.05–5%	Detergent
Tween-20	0.76	0.005–10%	Detergent
Tween-60	0.08	0.50%	Detergent
Tween-80	0.08	0.20%	Detergent

3. Methods

In this section, we first describe briefly how experimental data is structured in the REFOLD database and how it can be retrieved, for example, when constructing a refolding strategy. After describing the most common refolding techniques, we focus on how the experimentalist can use the REFOLD resource to either modify generic protocols or design new refolding protocols based upon properties of the target protein. In particular, we focus on key parts of the refolding strategy that are likely to determine outcome, and thus how the data in REFOLD can facilitate the protocol design process.

3.1. Refolding Data Structure

Refolding data is presented to the user via a simple one-page report, entailing details about the protein of interest as well as the refolding and purification procedures. This form is logically structured, such that properties of the protein are detailed first, followed by details of expression and finally the refolding methodology. This allows for a standardized format, and thus provides a streamlined reference catalogue.

For each entry, basic details regarding properties of the protein, such as *chain length, pI, molecularity, disulphide bonds, molecular weight,* and *species* are recorded. This part of the form also provides a cross-reference to the *UniProt* (15) and *SCOP* (16) databases and the *SCOP family* to which the protein belongs (if known). The entry of protein traits is then followed by details of the paper in which the protocol was originally published, with the *journal* name, paper *title*, publication details, and *PubMed* cross-reference

Details of protein expression comprise information, such as the *cell type* (bacterial, yeast, insect, etc.) and *strain,* in which the protein is expressed, as well as the expression *vector* used to encode the protein. The *cell density* at which protein expression is induced, as measured by optical density at 600 nm (OD_{600}), is also captured, as well as the *time* and *temperature* of expression.

The refolding protocol is one of the central aspects of the database. The report provides the details of the *refolding method,* i.e. the technique used to refold the protein, as well as various buffer conditions used in the protocol. This includes the *solubilization buffer* in which the protein is unfolded, the *wash buffer* that is used to wash inclusion bodies and remove cellular debris and loosely bound proteins, and finally the *refolding buffer* in which the protein is refolded. Details regarding refolding conditions, including *time, temperature, pH, redox reagents,* and *chaperones* (if used), are also specified, as well as other variables, such as *pre-purification* steps prior to refolding, *refolding yield* and *purity.* Importantly, comprehensive descriptions of the *protocol* for expression, refolding, and purification, as would be detailed in a paper, are provided.

3.2. Generic Refolding Protocols

A schematic outline of the procedure for protein refolding is shown in Fig. 1 and one for the isolation of inclusion bodies in Fig. 2. The following are generalized protocols for some of the most common procedures in protein refolding and are intended as basic starting points for the development of refolding protocols. Various parameters, such as buffer composition, volumes, pH, times, and temperatures will most likely need to be adjusted according to the specific properties of individual proteins in order to optimize the yield and efficiency of refolding (see Subheading 3.3).

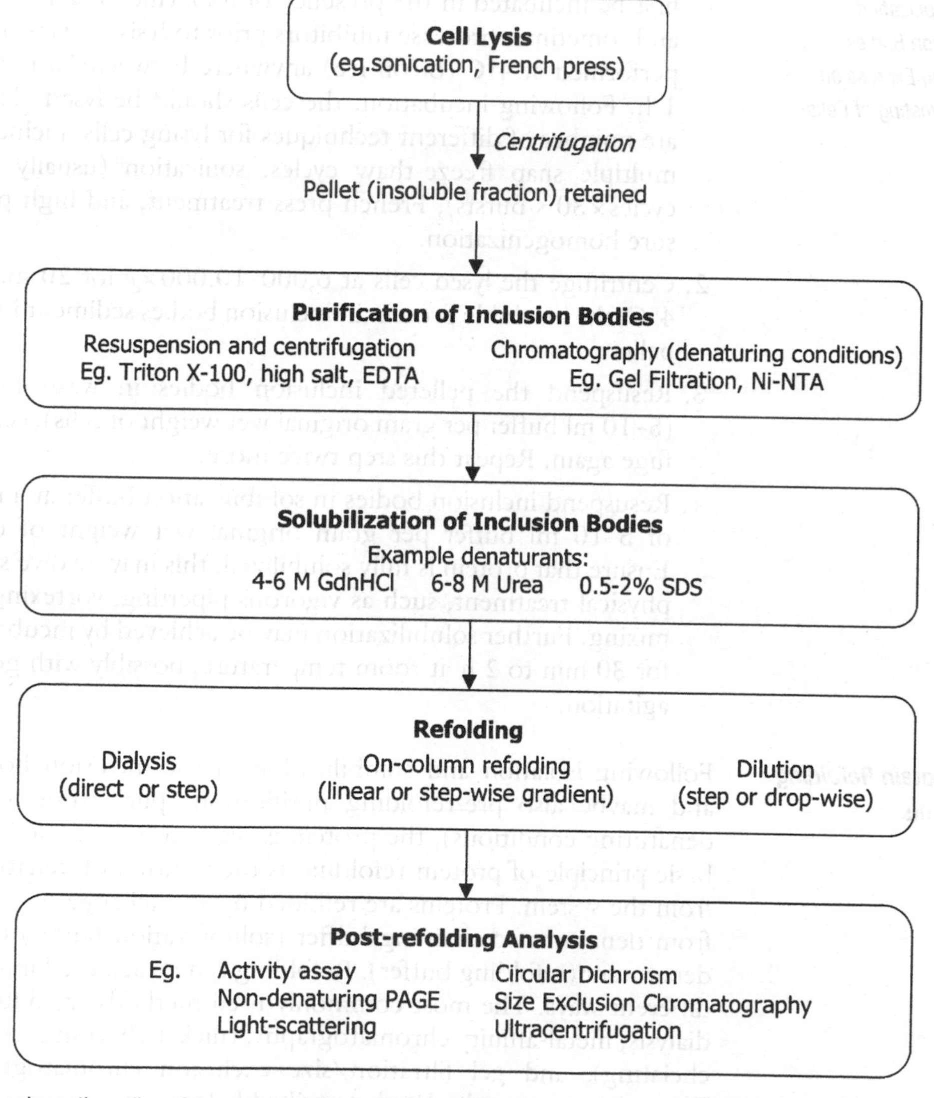

Fig. 1. A schematic outline of the procedure for purification and refolding of proteins from inclusion bodies.

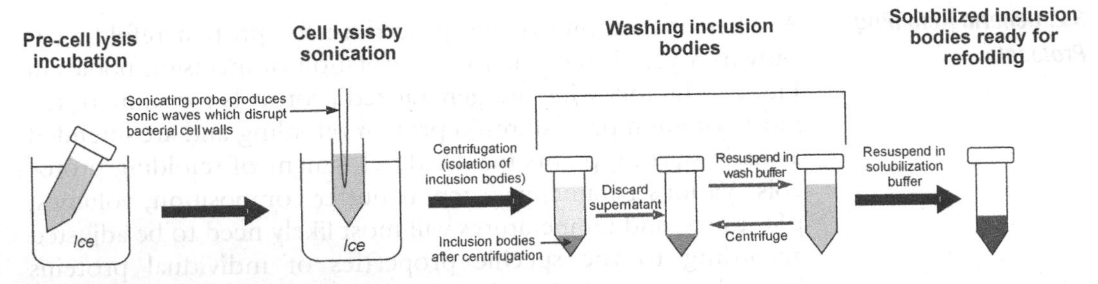

Fig. 2. Steps involved in the isolation and purification of inclusion bodies.

3.2.1. Isolation and Solubilization of Inclusion Bodies (Following Expression and Harvesting of Cells)

1. Cells may be lysed in a number of different ways. The cells may first be incubated in the presence of lysozyme (1–2 mg/ml) and sometimes protease inhibitors prior to lysis. This is often performed at 4°C (or on ice) anywhere between 15 min to 1 h. Following incubation, the cells should be lysed. There are number of different techniques for lysing cells, including multiple snap freeze-thaw cycles, sonication (usually 5–6 cycles × 30 s bursts), French press treatment, and high pressure homogenization.
2. Centrifuge the lysed cells at 6,000–10,000 × *g* for 20 min at 4°C, the insoluble protein in inclusion bodies sediment in the pellet.
3. Resuspend the pelleted inclusion bodies in wash buffer (5–10 ml buffer per gram original wet weight of cells), centrifuge again. Repeat this step twice more.
4. Resuspend inclusion bodies in solubilization buffer at a ratio of 5–10 ml buffer per gram original wet weight of cells. Ensure that protein is fully solubilized; this may involve some physical treatment, such as vigorous pipetting, vortexing, or mixing. Further solubilization may be achieved by incubation for 30 min to 2 h at room temperature, possibly with gentle agitation.

3.3. Protein Refolding Protocols

Following isolation and solubilization of the inclusion bodies, and maybe also pre-refolding purification (purification under denaturing conditions), the protein is ready to be refolded. The basic principle of protein refolding is the removal of denaturant from the system. Proteins are refolded by an exchange of buffers from denaturant-containing buffer (solubilization buffer) to no denaturant (refolding buffer). Refolding can be achieved in many different ways. The most commonly used methods are dilution, dialysis, metal-affinity chromatography, (nickel-chelating, cobalt-chelating), and gel filtration/size exclusion chromatography. These common methods are described below.

3.3.1. Dilution

1. Add the solubilized, unfolded protein (in denaturant) to 20–100× volume refolding buffer.
2. Leave the refolding mixture for at least 4–16 h at 4°C with gentle stirring.
3. Centrifuge or filter refolded protein mixture to remove any insoluble material that may have formed.
4. This simple method enables the protein to refold by diluting out the denaturant that is present. The dilution of the protein in a large volume of refolding buffer is also conducive to correct folding of the protein in a less crowded milieu. However, this results in the final protein concentration being relatively low, and the protein will most likely need to be concentrated after refolding. The final protein concentration is generally about 0.1–0.2 mg/ml or less (see Note 2).

3.3.2. Dialysis

1. Dialyze the solubilized, denatured protein against 40–100× volume of refolding buffer for at least 8 h at 4°C with stirring.
2. Remove refolded protein from dialysis tubing, centrifuge or filter refolded protein mixture to remove any insoluble material that may have formed.

This protocol allows for the gradual dilution of the denaturant in a large of volume of buffer without reducing the protein concentration. On the other hand, this may not be suitable for proteins that do not refold well at high concentrations. Typical protein concentrations in dialysis tubing are generally 0.1–0.5 mg/ml (see Note 3).

3.3.3. Column Refolding: Nickel-Chelating Chromatography

The method described here refers to nickel-chelating chromatography; however, the method for cobalt-chelating chromatography is essentially the same.

1. Equilibrate a nickel-chelated resin with solubilization buffer (containing denaturant) (see Note 4).
2. Apply the solubilized protein to the column. In order to maximize binding of the protein, the protein should be loaded slowly. If using loose resin, allowed to mix with the resin for at least 2 h. Proteins lacking the affinity tag do not bind the column and are eluted (see Note 5).
3. Wash the resin with 5–10 column volumes of solubilization buffer.
4. Refold the protein by applying a buffered gradient of decreasing denaturant concentration to the column (e.g. 8 M to 0 M urea). This may be a linear gradient with a gradual changeover of high-denaturant to low-denaturant buffers applied using a pump, or a step-wise gradient, involving a series of individual buffers each containing an incrementally reduced denaturant concentration (e.g. 6, 4, 2, and 0 M guanidine).

5. Elute the protein with refolding buffer containing a higher concentration of imidazole (e.g. 250–500 mM).

This method is suitable for proteins expressed with a polyhistidine tag, and may be adapted very similarly to column refolding using other agarose-linked metals (e.g. cobalt). While this is a little more complex than dilution and dialysis methods, column-assisted refolding has the advantage that the proteins are less likely to aggregate during refolding.

3.3.4. Column Refolding: Gel Filtration/Size-Exclusion Chromatography

1. Equilibrate the size-exclusion column of choice with refolding buffer.
2. Apply the denatured protein (in solubilization buffer) to the column.
3. Continue application of refolding buffer to the column. As the protein moves down the column, the buffer is exchanged, the denaturant is diluted out and the protein refolds and elutes.

This method allows the protein to refold as it passes through the gel filtration column. This also allows the removal of any large molecular weight aggregated species that may form during misfolding. However, the overall volume of the sample is increased and the final concentration of the protein reduced.

3.4. Designing a Refolding Protocol Using REFOLD

Typically, the strategy for refolding a protein starts by simply following one of the generic protocols described above. However, characteristics of the protein (molecular weight, number of disulphide bonds, as well as structural characteristics such as family) may guide modification of the protocol or even identification of new protocols, for example, based upon homology with other proteins. This is where the ability to search the structured data in REFOLD becomes extremely useful. Users can easily access data in the REFOLD database by executing a simple search on any chosen term, or alternatively, an advanced search can be performed according to more specific parameters. The tabulated search results provide details of a number of sortable parameters, such as various protein properties, family, and refolding method and conditions. Specified links provide access to full refolding records, while selecting the name of a protein leads to more detailed information about the protein itself. Additionally, following links from the search results page to other parameters lead to refolding records sharing that property. A *PubMed* cross reference also allows users access to the original article. Although the search is comprehensive, the following search parameters are most likely to help the user in protocol design: protein name, family, number of disulphides, isoelectric point, fusion tag, refolding technique, refolding additive, chaperones, and redox agents. The provision of a graphical breakdown of various parameters for refolding techniques and protein properties allows particular trends and patterns to be observed at a glance, providing information about the most common features and methods used.

3.4.1. Finding Refolding Protocols of Homologues: Using Protein Sequence Similarity

The majority of REFOLD entries contain the sequence of the refolded protein, allowing sequence similarities between entries to be pre-computed on a daily basis using BLAST (17). Thus, refolding protocols for proteins with similar sequences can be identified by simply entering the sequence into the Web page. In cases of significant sequence similarity for REFOLD entries, it is likely that refolding protocols will be appropriate, or at least a good starting point. Comparison of identified protocols indicates the parameters that can be varied.

3.4.2. The Refolding Protocol

In the following section, we highlight a selection of important criteria worth considering when designing a refolding protocol, together with some REFOLD statistics that give a good indication of the wide spectrum of experimental conditions in reported refolding methodologies.

The refolding buffer: The cocktail of pH buffer, salts, redox agents, and additives have a strong influence on the outcome of refolding. More than half of the entries in REFOLD report Tris–HCl as the buffer at pH range 7–9, followed by sodium phosphate (6.1%), PBS (5.3%), Hepes (4.6%), potassium phosphate (3.4%), and ammonium acetate (2.0%).

Folding aids: Although 32 different types of refolding additives have been reported in REFOLD, their use is uncommon. Common additives include detergents, polar additives, weak chaotrophs, osmolytes, and cations (summarized in Table 1). Additives act by either stabilizing the native or intermediate states on the folding pathway, or by preventing aggregation. A number of proteins have been refolded in the presence of additives, such as arginine (176 entries) and glycerol (146 entries), which are both compounds commonly used to aid the folding of proteins (6, 7, 11, 18), and the most common additives in REFOLD. In cases where arginine is used, it is generally present in the refolding buffer at concentrations ranging from 0.2 to 1.0 M, and the buffer pH ranges from 7.2 to 11.3, where the average pH is 8.3. In contrast, glycerol, which is generally used at concentrations between 1 and 50% (v/v), has been included in buffers with pH values ranging from 4.5 to 10.8 (average: 7.8). Other additives that are known to assist protein refolding have been used in a few instances, for example, ethylene glycol (5–50% v/v), Triton X-100 (0.05–5% v/v), and glycine (50 mM to 1.0 M). If the target protein is prone to proteolysis, protease inhibitors are included in the refolding buffer, the most common being PMSF (52 entries), and much less commonly combinations of aprotinin, leupeptin, and pepstatin reported. Molecular chaperones have been employed in only a few cases (2.8%), with the most common being GroEL and GroEL minichaperone (19).

Disulphide bond formation: A redox system is necessary for the correct folding of proteins containing disulphide bonds. This is typically achieved using a combination of the oxidized and reduced forms of redox agents, in equimolar to 10:1 ratios. The two most common redox systems are GSH/GSSG (17%), followed by DTT (15%), Beta-mercaptoethanol (8%), cysteamine/cystamine (4%), and Cysteine/Cystine (1%). Other combinations of redox agents have been reported, for example, Beta-mercaptoethanol/GSH-GSSG, Beta-mercaptoethanol/Cystamine, DTT/cysteine/cystine, and GSH/GSSG/DTT, though much less common than the above.

Refolding methods: Protocols for the most common refolding methods have been described in Subheading 3.3, but some comments on their prevalence are made here. Despite more that 41 refolding techniques reported in REFOLD, the vast majority use one of the four methods outlined in Subheading 3.3. By popularity, they rank dilution (40%); dialysis (27%); dilution/dialysis (11%); Nickel-chelating chromatography (5%); and size-exclusion chromatography (4%). These trends suggest that in most cases the simplest methods may be sufficient to yield adequate quantities of protein without the need to complicate the protocol further.

Folding/Fusion tags: Despite more than 80 fusion tags reported in REFOLD, the majority (60%) of proteins are expressed without a fusion tag. Of the proteins that are expressed with fusion tags, the most common tag is a N- (17%), followed by C-terminal hexahistidine (his_6) tag (7%), consistent with the fact that column-assisted refolding on a nickel-chelating resin is the third-most common refolding method. Common fusion tags are listed in Table 2.

3.4.3. Analysis of Refolded Protein

This quality control final step is required to measure the success of refolding, i.e. the presence of folded protein versus misfolded aggregate. If disulphides are present in the properly folded state,

Table 2
The most common fusion tags in the REFOLD database

Fusion tag	Size	Location
Hexahistidine	840 Da	N-terminus (20.3%) and C-terminus (8.8%)
Glutathione-S-transferase (GST)	26 kDa	N-terminus (1.29%) and C-terminus (0.1%)
Thioredoxin	11 kDa	N-terminus (0.7%)
Maltose-binding protein (MBP)	40 kDa	N-terminus (0.4%) C-terminus (0.2%)

it is also necessary to confirm that they have formed correctly. The "gold standard" analysis for enzymes is the activity assay, as this is the most sensitive measure of their proper function. Indeed, this the most common assay in the REFOLD database (31%). However, many proteins have no known structure or function, and this is consistent with more than 80 assays reported in REFOLD. Aside from activity, SDS-PAGE is the most popular post refolding analysis (13%), but this reflects its widespread use to determine purity and because it is conducted under denaturing conditions does not indicate correct folding. Far UV circular dichroism is the next most popular (6%), indicating its utility in distinguishing the α/β structure of folded proteins from the random coil of aggregates. Fluorescence techniques, including dynamic light scattering, are also powerful ways of detecting aggregates.

4. Notes

1. *Alternative cell lysis method*: Cells may be lysed and inclusion bodies unfolded simultaneously by the resuspension of harvested cells directly into solubilization buffer. The denaturant in the buffer serves to break down the bacterial cell wall. Cells should be incubated in solubilization buffer for 1–2 h at room temperature with gentle agitation.
2. *Possible variation on the basic protocol*: adding the solubilized protein slowly (by slow drip, approximately 0.1–0.2 ml/min) to the refolding buffer may increase the efficiency and yield of refolding by allowing individual protein slowly and individually, rather than all at once – this may be useful for proteins which are prone to aggregation from a partially folded intermediate conformation.
3. *Possible variation on the basic protocol–Step-wise dialysis*: Change dialysis buffers regularly (e.g. every 2–4 h), each time incrementally reducing the concentration of denaturant present (e.g. 6, 4, 2, and 0 M guanidine). This may be more suitable for some proteins that refold better over a gradual dilution of denaturant, but not for proteins that may aggregate from partially folded species present at intermediate denaturant concentrations.
4. A low concentration of imidazole (e.g. 5–10 mM) could be included in the solubilization buffer when equilibrating the resin, to reduce non-specific binding of proteins to the resin.
5. During the resin-washing stage, it may also be useful to include slightly higher concentration of imidazole (e.g. 5–20 mM). It is important to note, however, that this is

dependent on the protein of interest being bound tightly enough to the column so as not to be washed off in low imidazole concentrations.

Acknowledgments

The authors would like to acknowledge the contribution of all the researchers whose published data has been entered into REFOLD. This work was supported by grants from the National Health and Medical Research Council, the Victorian State Government, and the Victorian Partnership for Advanced Computing. SPB is a Monash University Senior Logan Fellow and NHMRC Senior Research Fellow. AMB is an NHMRC Senior Research Fellow.

References

1. Sorensen, H. P., and Mortensen, K. K. (2005) Soluble expression of recombinant proteins in the cytoplasm of Escherichia coli *Microb. Cell Fact.* **4**, 1.
2. Sorensen, H. P., and Mortensen, K. K. (2005) Advanced genetic strategies for recombinant protein expression in Escherichia coli *J. Biotechnol.* **115**, 113–28.
3. De Marco, V., Stier, G., Blandin, S., and de Marco, A. (2004) The solubility and stability of recombinant proteins are increased by their fusion to NusA *Biochem. Biophys. Res. Commun.* **322**, 766–71.
4. Dyson, M. R., Shadbolt, S. P., Vincent, K. J., Perera, R. L., and McCafferty, J. (2004) Production of soluble mammalian proteins in Escherichia coli: identification of protein features that correlate with successful expression *BMC Biotechnol.* **4**, 32.
5. Kapust, R. B., and Waugh, D. S. (1999) Escherichia coli maltose-binding protein is uncommonly effective at promoting the solubility of polypeptides to which it is fused *Protein Sci.* **8**, 1668–74.
6. Cabrita, L. D., and Bottomley, S. P. (2004) Protein expression and refolding - A practical guide to getting the most out of inclusion bodies *Biotechnol. Annu. Rev.* **10**, 31–50.
7. Fahnert, B., Lilie, H., and Neubauer, P. (2004) Inclusion bodies: formation and utilisation *Adv. Biochem. Eng. Biotechnol.* **89**, 93–142.
8. Middelberg, A. P. (2002) Preparative protein refolding *Trends Biotechnol.* **20**, 437–43.
9. Panda, A. K. (2003) Bioprocessing of therapeutic proteins from the inclusion bodies of Escherichia coli *Adv. Biochem. Eng. Biotechnol.* **85**, 43–93.
10. Tsumoto, K., Ejima, D., Kumagai, I., and Arakawa, T. (2003) Practical considerations in refolding proteins from inclusion bodies *Protein Expr Purif* **28**, 1–8.
11. Clark, E. D. B. (1998) Refolding of recombinant proteins *Curr. Opin. Biotechnol.* **9**, 157–63.
12. Buckle, A. M., Devlin, G. L., Jodun, R. A., Fulton, K. F., Faux, N., Whisstock, J. C., and Bottomley, S. P. (2005) The matrix refolded *Nat Methods* **2**, 3.
13. Chow, M. K., Amin, A. A., Fulton, K. F., Fernando, T., Kamau, L., Batty, C., Louca, M., Ho, S., Whisstock, J. C., Bottomley, S. P., and Buckle, A. M. (2006) The REFOLD database: a tool for the optimization of protein expression and refolding *Nucleic Acids Res* **34**, D207–12.
14. Chow, M. K., Amin, A. A., Fulton, K. F., Whisstock, J. C., Buckle, A. M., and Bottomley, S. P. (2006) REFOLD: an analytical database of protein refolding methods *Protein Expr Purif* **46**, 166–71.
15. Apweiler, R., Bairoch, A., Wu, C. H., Barker, W. C., Boeckmann, B., Ferro, S., Gasteiger, E., Huang, H., Lopez, R., Magrane, M., Martin, M. J., Natale, D. A., O'Donovan, C., Redaschi, N., and Yeh, L. S. (2004) UniProt: the Universal Protein knowledgebase *Nucleic Acids Res.* **32**, D115–9.

16. Murzin, A. G., Brenner, S. E., Hubbard, T., and Chothia, C. (1995) SCOP: a structural classification of proteins database for the investigation of sequences and structures *J. Mol. Biol.* **247**, 536–40.
17. Altschul, S. F., Madden, T. L., Schaffer, A. A., Zhang, J., Zhang, Z., Miller, W., and Lipman, D. J. (1997) Gapped BLAST and PSI-BLAST: a new generation of protein database search programs *Nucleic Acids Res* **25**, 3389–402.
18. Tsumoto, K., Umetsu, M., Kumagai, I., Ejima, D., Philo, J. S., and Arakawa, T. (2004) Role of arginine in protein refolding, solubilization, and purification *Biotechnol Prog* **20**, 1301–8.
19. Altamirano, M. M., Golbik, R., Zahn, R., Buckle, A. M., and Fersht, A. R. (1997) Refolding chromatography with immobilized mini-chaperones *Proc Natl Acad Sci U S A* **94**, 3576–8.

Chapter 5

Circular Dichroism and Its Use in Protein-Folding Studies

David T. Clarke

Abstract

The way in which proteins fold into the complex 3 dimensional structures that are responsible for their function is a subject of great practical as well as fundamental significance because of the involvement of folding and misfolding in a number of serious human and animal diseases. Ultraviolet circular dichroism (CD) reports on the secondary and tertiary structure of proteins. Measurements can be made on proteins in the solution phase, and critically time-resolved measurements can be made with millisecond resolution. This combination of characteristics makes CD a useful tool for investigating protein folding, and indeed any process involving changes in protein structure. Experimental methods for a typical time-resolved CD experiment are described, and some common problems identified.

Key words: Protein folding, Circular dichroism, Stopped-flow, Ultraviolet spectroscopy, Protein secondary and tertiary structure

1. Introduction

The folding of proteins to their predetermined native conformations is a fundamental process in biology with enormous practical implications (1). For example, in the biotechnology industry, misfolded overexpressed proteins frequently accumulate in inclusion bodies, resulting in expensive loss of product. Also, a growing number of diseases are attributed to the reorganisation of secondary structure (the so-called misfolding). Key phases of protein folding are the development of secondary structure and the changes in overall molecular dimensions that accompany the folding process (2). A number of techniques have been employed to study the folding process, and one that has found extensive application is ultraviolet circular dichroism (CD).

Andrew F. Hill et al. (eds.), *Protein Folding, Misfolding, and Disease: Methods and Protocols*, Methods in Molecular Biology, vol. 752, DOI 10.1007/978-1-60327-223-0_5, © Springer Science+Business Media, LLC 2011

CD is a form of optical activity that arises from the differential absorption of left and right circularly polarised light by a solution of chiral molecules (3, 4). CD measurements have for many years been employed in the investigation of protein structure. In the "far UV" (wavelengths between 180 and 260 nm), the CD of proteins arises because of the amide chromophore of the peptide bond. Although the amide chromophore itself has no optical activity, the presence of the asymmetric α-carbon atoms on either side of the bond induces a CD signal, and this signal reports on the secondary structure of the protein (5). In the "near UV" (wavelengths between 250 and 350 nm), the chromophores are the aromatic amino acids and disulphide bonds, and the CD is sensitive to the tertiary structure of the protein (6). Specifically, signals in the 250–270 nm region arise from phenylalanine, signals from 270 to 290 nm are due to tyrosines, and from 280 to 300 nm signals are from tryptophan. Disulphide bonds give weaker signals throughout the near UV region.

CD spectrometers are available from a number of manufacturers, and use xenon arc lamps as their light source (7). More recently, synchrotron radiation has also been used as a light source for CD, and has certain advantages, such as increased brightness and extended wavelength range (8, 9). The methods described below should be adaptable to all commonly available CD instruments.

CD has been used extensively to investigate dynamic processes in proteins, particularly protein folding (10). CD readily lends itself to these types of studies for a number of reasons. Firstly, it provides information on structure, secondly, measurements are made in the solution phase, allowing the easy initiation of dynamic processes, and thirdly, the time resolution of the technique (millisecond) is well matched to many processes of interest (11). Time-resolved CD measurements can be made in the far UV to investigate secondary structure formation or in the near UV to study the development of tertiary structure. The most common method of initiating folding or unfolding of proteins in time-resolved CD studies is by stopped-flow (12). For folding studies, proteins are typically mixed with denaturants, such as urea or guanidine hydrochloride, and then rapidly mixed with a diluent to reduce denaturant concentration to a level at which folding can occur. The time resolution of stopped-flow measurements is limited by the dead time of the apparatus, which is effectively the time after mixing during which the reaction cannot be monitored. Dead time depends on the flow rate and volumes of the sample and diluent, and is typically of the order of 0.5–2.0 ms (13). There are a number of stopped-flow devices available that are suitable for CD measurements. The methods described below are not specific to a particular stopped-flow apparatus, but should be applicable to all. Methods other than stopped-flow, such as laser temperature-jump (14), have also been used to initiate protein folding and unfolding for CD studies. To date, these techniques require specialised apparatus and are not addressed in this chapter.

The chapter is written with the aim of providing instructions for the setting-up and carrying out of a stopped-flow protein-folding CD experiment. However, the sections on sample characterisation by steady-state CD, and on CD data analysis, are equally applicable to the characterisation of any protein by CD for the determination of secondary structure content.

2. Materials

2.1. Sample Characterisation by Steady-State CD Spectroscopy

1. CD sample cells (Hellma) (see Note 1).
2. Sample buffer, e.g. 20 mM sodium phosphate, pH 7.0 (see Note 2).
3. Denaturant solution at a range of concentrations (e.g. 0.5–6.0 M). Urea or Guanidine Hydrochloride is the most commonly used (see Note 3).

2.2. Stopped-Flow Calibration

1. 0.5 mM 2,6-dichloroindophenol (DCIP) (Sigma Aldrich). Firstly, make up a stock solution of 2 mM DCIP in 0.1 M sodium chloride, and then dilute to a final DCIP concentration of 0.5 mM with 0.1 M sodium acetate, pH 5.0. The DCIP solutions have a short shelf life and should be made up fresh as required.
2. 100 mM ascorbic acid, pH 5.0. Make up by dissolving ascorbic acid in 0.1 M NaCl, and adjust the pH to 5.0 using concentrated sodium hydroxide solution. Dilute as appropriate to obtain an ascorbic acid concentration series between 5 and 25 mM. Again, the ascorbic acid solutions should be made up fresh for each calibration.

2.3. Stopped-Flow CD Experiment

1. Sample dilution buffer. This varies depending on the requirements of the sample, as with the steady-state measurement (see Note 2).
2. Protein sample dissolved in denaturant at an appropriate concentration (for determination of appropriate denaturant concentration see Subheading 3.2). For protein concentration, see Note 4.
3. Protein dissolved in dilution buffer at the same protein concentration as the protein sample in denaturant.

3. Methods

3.1. Sample Characterisation by Steady-State CD Spectroscopy

Characterisation of the protein under investigation in the folded and unfolded states is the key to a successful stopped-flow CD experiment. Obtaining steady-state CD spectra allows the experimenter to determine the feasibility of the proposed experiment, select appropriate wavelength(s) for the time-resolved work, and analyse the data. The steps outlined below assume that the CD instrument has already been calibrated.

1. Switch on the CD spectrometer and allow time for the electronics and lamp to stabilise (approximately 30 min). Pay particular attention to requirements for nitrogen purging as this is important to avoid contamination of the optics.
2. Set the sample holder to an appropriate temperature (see Note 5).
3. Load an appropriate sample cell with a solution of protein in buffer (see Note 6). Protein concentrations should preferably be in the range from 1.0 to 10.0 mg/ml.
4. Place the cell in the CD instrument and collect a CD spectrum. The achievable wavelength range of the spectrum depends on the instrument, the solvent, and the cell pathlength, as well as the type of information sought. Where the requirement is to investigate secondary structure, spectra should be collected in the far UV between 180 and 260 nm. For tertiary structure investigations, near UV wavelength ranges between 260 and 350 nm are appropriate. Scanning times again depends on the type of instrument but typical data collection times for a spectrum are between 10 and 30 min.
5. Repeat step 4 as required and average spectra to obtain a single spectrum with sufficiently high signal-to-noise.
6. Repeat steps 3–5, loading the cell with the buffer in which the protein sample was dissolved.
7. Repeat steps 3–6 with samples at a range of denaturant concentrations.

3.2. Assessment of Conditions for Stopped-Flow Experiment

This section describes basic analysis and use of the data obtained in Subheading 3.1 to determine the feasibility of, and appropriate conditions for, the stopped-flow experiment.

1. Average all spectra collected for protein in each buffer and denaturant concentration, and also buffer spectra.
2. Subtract the appropriate buffer spectrum from the protein spectrum to obtain the CD spectrum for the protein (see Note 7).

3. Depending on the instrument used, it may be necessary to convert the units of the spectra. As an example, the data may be output in ellipticity (millidegrees), and it is necessary to convert to mean residue ellipticity for data analysis. Most commonly, the protein concentration is known in mg/ml and the following equation can be applied:

$$[\theta] = \theta / (c / 11) l$$

Where $[\theta]$ is the mean residue ellipticity, θ is the ellipticity, c is the sample concentration in mg/ml, and l is the cell pathlength in cm (15).

4. Plot the spectra. A typical set of spectra is shown in Fig. 1. The spectra should show a clear transition from that typical of unfolded protein at high denaturant concentration (strong negative CD band around 200 nm) to the spectrum of a folded protein (e.g. negative CD bands around 208 and 222 nm for a helical protein) at low denaturant concentration.
5. Using the spectra, determine the denaturant concentration at which the protein is fully unfolded, and the concentration at which it is fully folded. This is easily done by locating the denaturant concentrations above and below which no spectral changes occur. For the example given in Fig. 1, the protein is fully unfolded in 5.0 M urea, and fully folded at a urea concentration of 1.0 M. This determines the starting denaturant concentration, and stopped-flow dilution required for a full transition from unfolded to folded protein.

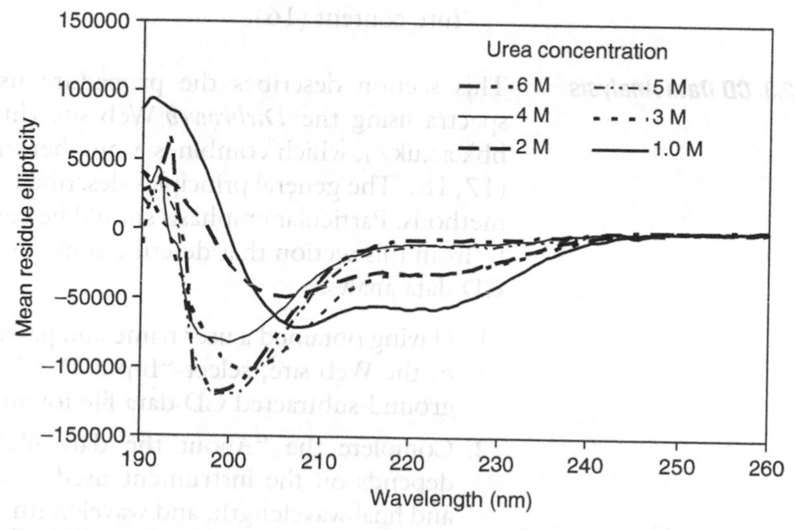

Fig. 1. CD spectra of apo-myoglobin taken at a range of urea concentrations, for the determination of appropriate conditions for stopped-flow folding experiment. Spectra were recorded on the CD12 SRCD beamline at the SRS, Daresbury Laboratory.

6. For a tertiary structure experiment, find the wavelength at which there is maximum difference between the signals for folded and unfolded protein. When this wavelength has been determined, it is possible to proceed to the stopped-flow measurements. It may be desirable to measure the kinetics at several wavelengths to investigate, for example, signals arising from tryptophan and disulphide bonds.
7. When the object of the experiment is to determine the kinetics of secondary structure development, further data analysis is required to determine the secondary structure content of the folded protein. This can be achieved with one of the several analysis programmes that are easily available. The procedure is described in Subheading 3.3.
8. The secondary structure content of the protein determined by following the procedure in Subheading 3.3 can be used to decide on the best approach for the stopped-flow experiment. For a protein containing only one secondary structure type, a single wavelength stopped-flow measurement gives meaningful kinetic information. Wavelengths of maximum change in CD signal can be identified from the spectra. For all-helix proteins, the best wavelength should be around 222 nm, and for all-beta proteins around 215 nm. Where the protein contains significant amounts of more than one secondary structure type, the situation is more complex. Multiple wavelengths should be selected to attempt to identify folding rates for different structure types. Ideally, given sufficient time and sample, measurements should be made at intervals throughout the CD spectrum, allowing the reconstruction of time-resolved CD spectra that can be analysed for secondary structure content (16).

3.3. CD Data Analysis

This section describes the procedure used for analysing CD spectra using the *Dichroweb* Web site (http://dichroweb.cryst.bbk.ac.uk/), which combines a number of CD analysis packages (17, 18). The general principles described apply to all CD analysis methods. Particular emphasis should be paid to the notes referred to from this section that describe some of the common pitfalls of CD data analysis.

1. Having obtained a user name and password for *Dichroweb*, go to the Web site, select “Input Data”, and select your background-subtracted CD data file for analysis.
2. Complete the “About the data file” section. File format depends on the instrument used to collect the data. Initial and final wavelength, and wavelength step should be obvious. In the box for “Lowest wavelength datapoint to use in the analysis” enter the shortest wavelength at which good data have been obtained (see Note 8).

3. Select an analysis method (see Note 9).
4. Select a reference set that matches the wavelength range of your data.
5. In the "Advanced options" section, the optional scaling factor should be left at the default value of 1.0 for the initial analysis. This scaling factor allows estimates to be made of the effect of protein concentration errors on the analysis (see Note 10).
6. Select your preferred output units and click the "submit" button. Then, check the information that is to be submitted, and if this is correct, click "continue."
7. The data is plotted in your browser window. If there are no problems with the spectrum, click "continue."
8. The results page then appears. "View summary" gives the secondary structure information. More detailed information on the analysis can be found under "All of it!" It is also advisable to check the "Graph it!" window, where the experimental data can be seen plotted with the reconstructed data from the reference set. If there are large differences between the two spectra, the results should be treated with caution, and it is advisable to return to step 1 and rerun the analysis using a different method.
9. Return to step 1 and repeat the analysis for the next protein spectrum.

3.4. Stopped-Flow Set-Up and Calibration

It is important to ensure that the stopped-flow apparatus is correctly set up and functioning well before starting the experiment. This section describes the procedure for determining the dead time of the apparatus using the reduction (and associated decolourisation) of DCIP by ascorbic acid (19). It is good practice to perform this procedure at regular intervals, to check that the stopped-flow is performing to specification. This is particularly important when measuring fast processes that are close to the time resolution limit of the technique (milliseconds). This procedure requires the measurement of absorption rather than CD, and this facility is available on most instruments.

1. Set up the stopped-flow device to work at the mixing ratio that is used in the folding experiment.
2. Fill one stopped-flow syringe or reservoir with 0.5 mM DCIP solution, and the other with 5.0 mM ascorbic acid.
3. Set the apparatus to measure absorption at 610 nm, and to collect data for around 200 ms.
4. Start data acquisition and collect as many stopped-flow shots as required to obtain good signal-to-noise data.
5. Repeat with a higher concentration of ascorbic acid.

6. Repeat for several more ascorbic acid concentrations in the range from 5 to 25 mM.
7. Plot log(absorbance) vs. time for each ascorbic acid concentration.
8. Obtain the slope of each line and plot these against the log of the initial absorbance observed in each measurement.
9. Obtain the slope of the plot obtained in step 9. This slope is equal to the dead time of the stopped-flow apparatus (see Note 11).

3.5. Collection of Stopped-Flow CD Data

This section describes the basic steps required to obtain stopped-flow CD data from a protein sample. Although, there are many possible experimental variations that cannot be discussed here, the steps described should be a useful guideline. Obviously, the operation of stopped-flow instruments from different manufacturers varies, and it is assumed that operators have familiarised themselves with the correct operating procedures. Particular attention should be paid to the control measurements described; CD stopped-flow measurements are by nature subject to a number of artefacts, and the necessary controls described in steps 2–10 must be performed to ensure that the data obtained are reliable.

1. Set up the stopped-flow apparatus at the wavelength and mixing ratio determined in Subheading 3.2.
2. Set an appropriate data collection time and time per data point.
3. *First control measurement.* Load the protein dissolved in dilution buffer into the appropriate stopped-flow syringe or reservoir, and fill the other reservoir with dilution buffer.
4. Prime the stopped-flow transfer lines and cell using the procedure recommended by the manufacturer.
5. Collect a series of stopped-flow shots, sufficient to obtain a good signal (see Note 12).
6. The stopped-flow trace should be flat, and the amplitude of the CD signal recorded should be the same as the amplitude recorded in the steady-state CD spectrum of the folded protein collected earlier (see Note 13).
7. *Second control measurement.* Load the protein dissolved in denaturant into the appropriate stopped-flow syringe or reservoir, and fill the other reservoir with denaturant.
8. Prime the stopped-flow transfer lines and cell using the procedure recommended by the manufacturer.
9. Collect a series of stopped-flow shots, sufficient to obtain a good signal (more shots are probably required than in Step 5 because of the high absorbance of the denaturant).

10. Again, the stopped-flow trace should be flat, and the amplitude of the CD signal recorded should be the same as the amplitude recorded in the steady-state CD spectrum of the unfolded protein collected earlier. If this is not the case, again see Note 13.
11. *Folding experiment.* Load the protein dissolved in denaturant into the appropriate stopped-flow syringe or reservoir, and fill the other reservoir with dilution buffer.
12. Prime the stopped-flow transfer lines and cell using the procedure recommended by the manufacturer.
13. Collect a series of stopped-flow shots, sufficient to obtain a good signal.
14. At this stage, it is necessary to check the stopped-flow data to determine whether an appropriate time scale and time resolution have been selected.
15. Ideally, the stopped-flow trace should show a transition from a signal equivalent to that seen for unfolded protein, to a signal indicating fully folded protein. However, on many occasions this is not observed. Figure 2 shows some possible stopped-flow traces with explanations and suggested actions to be taken.
16. If necessary, adjust the stopped-flow data collection parameters according to the suggestions shown in Fig. 2, and repeat the experiment.
17. Repeat the procedure for any other wavelengths of interest.
18. Analysis of stopped-flow data can be made using a range of kinetics software packages. Suitable software is often provided with the stopped-flow apparatus. An example of kinetic fitting of stopped-flow data is shown in (16).

4. Notes

1. Good quality CD spectra, particularly those in the far UV region between 180 and 260 nm, require the use of cells with relatively short pathlengths. This is to overcome absorption of the short wavelength ultraviolet light by components of the sample buffer. This is particularly important when attempting to measure CD spectra in the presence of concentrated denaturants. Two types of cells are available; "bottle" type cells which are sealed units with filling holes, and demountable cells that consist of two plates between which the sample is sandwiched. For pathlengths of less than 0.5 mm, demountable cells are recommended as the bottle cells can be difficult to fill.

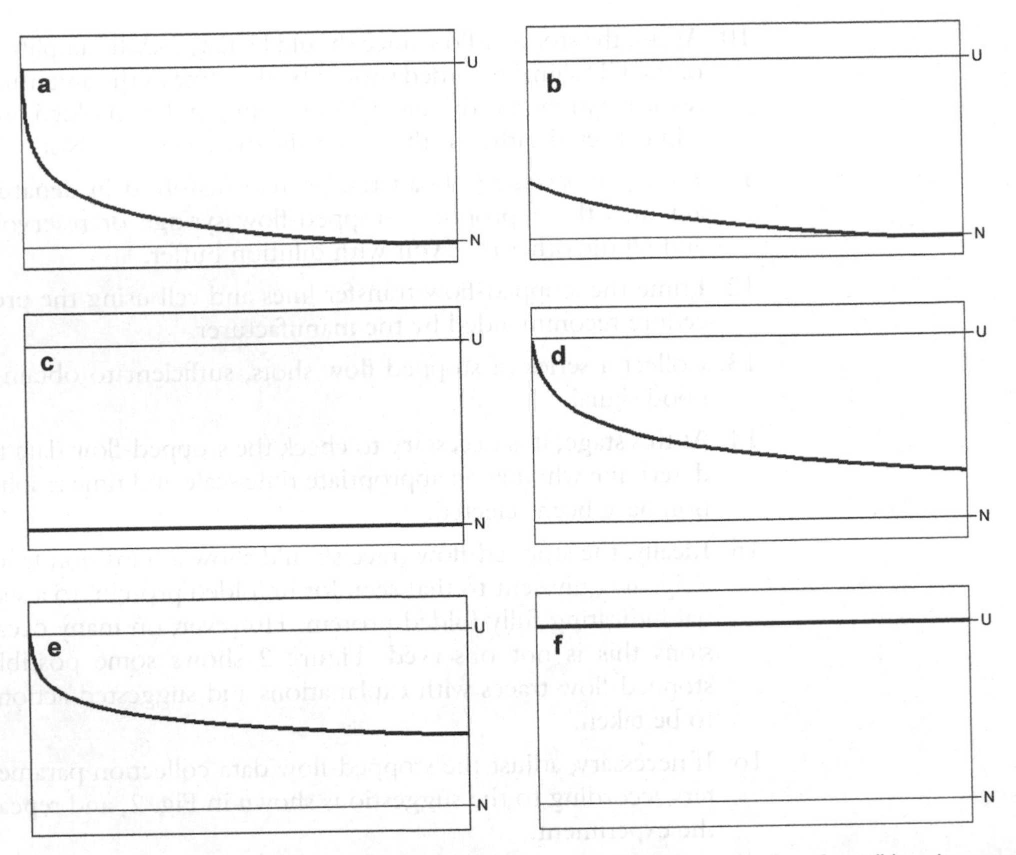

Fig. 2. Simulated stopped-flow traces (*bold lines*, CD signal vs. time) illustrating a number of possible outcomes of a typical folding experiment. Lines labelled "U" and "N" represent CD signals for unfolded and native protein, respectively. These values can be obtained from steady-state CD spectra (example shown in Fig. 1). (**a**) Stopped-flow parameters have been set correctly and the entire folding process has been monitored. (**b**) The initial part of the folding process has been missed. If possible, collect data again with faster time resolution. It is possible that the initial phase of folding is faster than the dead-time of the apparatus, in which the beginning of the folding process cannot be monitored by stopped-flow CD. (**c**) The protein is entirely folded within the dead time of the measurement. If the denaturant concentration is correct, and similar traces are obtained at the fastest achievable time resolution, the process is too fast to be monitored using stopped-flow CD. (**d**) The folding process is not complete within the time of the measurement. Increase the time of data collection and repeat. In some cases, fast and slow phases are observed; in this case, two different measurements on different time scales may be required to accurately determine the kinetics of each phase. (**e**) The process is complete, but the protein has not fully folded. Check denaturant concentration (may be too high) and stopped-flow mixing ratio (possibly insufficient dilution of denaturant). (**f**) Folding has not occurred. Again, check denaturant concentration and stopped-flow mixing ratio. Check that both sample and reagent syringes are being driven correctly.

2. The exact choice of buffer depends on the requirements of the protein being investigated. Buffer components should be minimised to avoid absorption problems discussed in Note 1. Relatively low concentration phosphate buffers are good. If the protein requires it, salts such as sodium chloride may be added but these will lower signal-to-noise, particularly at short wavelengths.

3. Generally, urea is the preferred denaturant as guanidine hydrochloride introduces ionic strength variation. Some proteins may not unfold in saturated urea solutions, and in this case guanidine hydrochloride or another alternative must be used. Whatever the denaturant, the high concentrations that are required mean that the highest possible grade reagent should be used.
4. Determining the correct protein concentration for stopped-flow experiments is to some extent a matter of trial and error, and depends on the wavelength of the measurement and the pathlength of the stopped-flow cell. For example, if the pathlength of the cell is 2 mm a suitable protein concentration after mixing of between 0.1 and 0.5 mg/ml is a good starting point for a far UV experiment. If a 10:1 stopped-flow dilution is to be used, the starting protein concentration should therefore be between 1 and 5 mg/ml. Higher concentrations are required for a near UV measurement.
5. The temperature selected should be the same as that which is used for the stopped-flow measurement. Folding rates are temperature dependent and there is no "correct" temperature to choose. The most commonly employed temperature employed in the literature is 25°C, and this has been suggested as a standard for folding kinetic studies (20).
6. A good rule of thumb for cell pathlength selection is to use the formula $p = 0.1/c$, where p is the pathlength in mm and c is the protein concentration in mg/ml. This applies to secondary structure CD measurements between 180 and 260 nm. This is a good place to start but can be adjusted depending on signal-to-noise. Measurements at longer wavelengths require pathlengths typically around 10–100 times longer. When loading the cells, it is important to ensure that no air bubbles remain trapped. Degassing the buffers and samples is advisable.
7. CD data are obtained in a number of data formats, depending on the instrument used. Basic data reduction procedures can be performed using standard spreadsheet software. Alternatively, a data processing package designed specifically for CD data, *cdtool*, is available at http://cdtools.cryst.bbk.ac.uk/ (21).
8. Selecting the lowest wavelength at which good data have been obtained is, to some extent, subjective. The noise level of CD data generally increases as the wavelength gets shorter because of increasing absorption of the light by the sample. The point at which the data goes bad is the point, where the signal is no longer discernible above the noise. This is typically characterised by sharp changes in CD signal. If in doubt, err on the longer wavelength side as the inclusion of bad data can adversely affect the analysis.

9. The analysis method that generally gives the most accurate results appears to be *cdsstr*. However, this method takes longer to run (up to 15 min while the others are almost instantaneous). If the time required for analysis is problematic, users may prefer to use one of the other methods, such as *selcon*, reserving *cdsstr* for spectra that other methods cannot analyse. If knowledge of the exact secondary structure content is critical for your experiments, it is good practice to analyse the same spectrum with a number of methods, to obtain an estimate of the degree of uncertainty in the results.
10. The effect of protein concentration errors on CD data analysis has been the subject of much discussion (22, 23). There is no doubt that relatively small protein concentration errors can have significant effects on the accuracy of the secondary structure content data obtained. However, the magnitude of the effects varies greatly from protein to protein, depending on the proportions of secondary structure types present. It is advisable to test the extent of the effects using the "optional scaling factor" in *Dichroweb*. For example, if it is believed that the protein concentration is accurate to within ±10%, repeat the analysis with the scaling factor set to 0.9, and again with it set to 1.1. This gives the variation in secondary structure content that would be expected from the maximum expected protein concentration error, and can be used to place confidence limits on the data.
11. The full dead time determination procedure does not have to be followed before each experiment. A quick check of the performance of the stopped-flow apparatus can be obtained by running a single ascorbic acid concentration. If the initial absorbance and slope are the same as previously observed, the apparatus is performing to specification.
12. The number of stopped-flow shots that have to be collected and averaged to obtain a good signal varies depending on the wavelength of the measurement and the absorption of the sample. This must be determined by trial experiments, and a judgement can be made on the basis of the steady-state CD spectra, which shows the magnitude of change in signal that is seen on folding/unfolding of the protein. Sufficient stopped-flow shots must be averaged so that the noise level of the data is significantly lower than the magnitude of the expected signal change.
13. Common problems that can be seen at this point are traces that slope, or have large spikes or ripples. There are a number of possible causes, including bubbles or large particles (e.g. protein aggregates) in the sample, or poor mixing due to a stopped-flow problem, such as a partially blocked sample line. These possible causes should be investigated, rectified, and

the measurement repeated. It is advisable to degas the sample and diluent if bubbles appear to be present in the stopped-flow cell. The other problem that might be detected at this stage is that amplitude of the CD signals measured in the stopped-flow and steady-state measurements are different. This can be caused by an incorrect stopped-flow mixing ratio, or a miscalibration of the stopped-flow apparatus. Again, these should be checked and rectified before continuing with the experiment.

Acknowledgements

The author would like to thank Prof. G.R. Jones and Mr. A. Brown who participated in the collection of some of the illustrative data shown. Data were collected on the CD12 beamline of the Synchrotron Radiation Source, Daresbury Laboratory, UK. The beamline was funded by the Biotechnology and Biological Sciences Research Council.

References

1. Dobson, C.M. (2000) in *Mechanisms of Protein Folding* (Pain, R.H., ed.) IRL Press, UK, pp. 1–33.
2. Dill, K.A., and Chen, H.S. (1997) From Levinthal to pathways to funnels. *Nature Struct. Biol.* **4**, 10–19.
3. Rodger, A., and Nordén, B. (1997) *Circular Dichroism and Linear Dichroism*, Oxford University Press, London and New York.
4. Jones, G.R., and Munro, I.H. (2000) in *Structure and Dynamics of Biomolecules* (Fanchon, E., ed.), Oxford University Press, London and New York.
5. Venyaminov, S.Y., and Yang, J.T. (1996) Determination of protein secondary structure, in *Circular Dichroism and the Conformational Analysis of Biomolecules* (Fasman, G.D., ed.), Plenum Press, New York and London, pp. 69–107.
6. Woody, R.W., and Dunker, A.K. (1996) Aromatic and cystine side-chain circular dichroism in proteins, in *Circular Dichroism and the Conformational Analysis of Biomolecules* (Fasman, G.D., ed.), Plenum Press, New York and London, pp. 109–157.
7. Johnson, W.C., Jr. (1996) Circular dichroism instrumentation, in *Circular Dichroism and the Conformational Analysis of Biomolecules* (Fasman, G.D., ed.), Plenum Press, New York and London, pp. 635–652.
8. Clarke, D.T., and Jones, G.R. (2004) CD12: a new high-flux beamline for ultraviolet and vacuum-ultraviolet circular dichroism on the SRS, Daresbury. *J. Synch. Rad.* **11**, 142–149.
9. Miles, A.J., Hoffman, S.V., Tao, Y., Janes, R.W., and Wallace, B.A. (2007) Synchrotron radiation circular dichroism (SRCD) spectroscopy: New beamlines and new applications in biology, *Spectroscopy* **21**, 245–255.
10. Nölting, B. (2005) in *Protein Folding Kinetics*, Springer, Berlin, Heidelberg, New York, pp. 98–104.
11. Roder, H., Elove, G.A., and Ramachandra Shastry, M.C. (2000) in *Mechanisms of Protein Folding* (Pain, R.H., ed.) IRL Press, UK, pp. 65–83.
12. Gibson, Q.H. (1969) Rapid mixing: Stopped flow. *Meth. Enzymol.* **16**, 187–228.
13. Hiromi, K. (2006) Recent Developments in the Stopped-Flow Method for the Study of Fast Reactions. In *Methods of Biochemical Analysis, Volume 26* (Glick, D. ed.), Wiley, New York, pp. 137–164.
14. Gai, F., Du, D., and Xu, Y. (2007) Infared Temperature-Jump Study of the Folding Dynamics of α-Helices and β-Hairpins. In

Protein Folding Protocols (Bai, Y., and Nussinov, R., eds.), Humana Press, New Jersey, pp. 1–20.

15. Woody, R.W. (1996) Theory of Circular Dichroism of Proteins, in *Circular Dichroism and the Conformational Analysis of Biomolecules* (Fasman, G.D., ed.), Plenum Press, New York and London, pp. 25–67.
16. Jones, G.R., and Clarke, D.T. (2004) Applications of extended ultra-violet circular dichroism spectroscopy in biology and medicine. *Faraday Discuss.* **126**, 223–236.
17. Whitmore, L. and Wallace, B.A. (2008) Protein secondary structure analyses from circular dichroism spectroscopy: methods and reference databases. *Biopolymers* **89**, 392–400.
18. Whitmore, L. and Wallace, B.A. (2004) DICHROWEB, an online server for protein secondary structure analyses from circular dichroism spectroscopic data. *Nucleic Acids Res.* **32**, W668-673.
19. Tonomura, B., Nakatani, H., Ohnishi, M., Yamaguchi-Ito, J., and Hiromi, K. (1978) Test reactions for a stopped-flow apparatus. *Anal. Biochem.* **84**, 370–383.
20. Maxwell, K.L., Wildes, D., Zarrine-Afsar, A., De Los Rios, M.A., Brown, A.G., Friel, C.T., Hedberg, L., Horng, J.-C., Bona, D., Miller, E.J., Vallée-Bélisle, A., Main, E.R.G., Bemporad, F., Qiu, L., Teilum, K., Vu, N.-D., Edwards, A.M., Ruczinski, I., Poulsen, F.M., Kragelund, B.B., Michnick, S.W., Chiti, F., Bai, Y., Hagen, S.J., Serrano, L., Oliveberg, M., Raleigh, D.P., Wittung-Stafshede, P., Radford, S.E., Jackson, S.E., Sosnick, T.R., Marqusee, S., Davidson, A.R., and Plaxco, K.W. (2005) Protein folding: Defining a "standard" set of experimental conditions and a preliminary kinetic data set of two-state proteins. *Protein Sci.* **14**, 602–616.
21. Lees, J.G., Smith, B.R., Wien, F., Miles, A. J., and Wallace, B.A. (2004) CDtool - An Integrated Software Package for Circular Dichroism Spectroscopic Data Processing, Analysis and Archiving. *Anal. Biochem.* **332**, 285–289.
22. McPhie, P. (2008) Concentration-independent estimation of protein secondary structure by circular dichroism: a comparison of methods. *Anal. Biochem.* **375**, 379–381.
23. Miles, A.J., Wien, F., Lees, J.G., and Wallace B.A. (2005) Calibration and standardisation of synchrotron radiation and conventional circular dichroism spectrometers. Part 2: factors affecting magnitude and wavelength. *Spectroscopy* **19**, 43–51.

Chapter 6

Distance Measurements by Continuous Wave EPR Spectroscopy to Monitor Protein Folding

James A. Cooke and Louise J. Brown

Abstract

Site-Directed Spin Labeling Electron Paramagnetic Resonance (SDSL-EPR) offers a powerful method for the structural analysis of protein folds. This method can be used to test and build secondary, tertiary, and quaternary structural models as well as measure protein conformational changes in solution. Insertion of two cysteine residues into the protein backbone using molecular biology methods and the subsequent labeling of the cysteine residues with a paramagnetic spin label enables the technique of EPR to be used as a molecular spectroscopic ruler. EPR measures the dipolar interaction between pairs of paramagnetic spin labels to yield internitroxide distances from which quantitative structural information on a protein fold can then be obtained. Interspin dipolar interaction between two spin labels at less than 25 Å are measured using continuous wave (CW) EPR methods. As for any low-resolution distance methods, the positioning of the spin labels and the number of distance constraints to be measured are dependent on the structural question being asked, thus a pattern approach for using distance sets to decipher structure mapping, including protein folds and conformational changes associated with biological activity, is essential. Practical guidelines and hints for the technique of SDSL-EPR are described in this chapter, including methods for spin labeling the protein backbone, CW-EPR data collection at physiological temperatures and two semi-quantitative analysis methods to extract interspin distance information from the CW-EPR spectra.

Key words: Site-directed spin labeling, Electron paramagnetic resonance, Protein folding, Continuous wave EPR spectroscopy, Site-directed mutagenesis, Interspin distances, Dipolar interaction

1. Introduction

Structural analysis of protein folds is most often performed using Nuclear Magnetic Resonance (NMR) spectroscopy or X-ray crystallography. However, large molecular weight proteins, protein complexes or proteins that are highly dynamic in solution are not always amenable to these high resolution techniques. Additionally, sufficient quantities of a protein can often be difficult to

Andrew F. Hill et al. (eds.), *Protein Folding, Misfolding, and Disease: Methods and Protocols*, Methods in Molecular Biology, vol. 752, DOI 10.1007/978-1-60327-223-0_6, © Springer Science+Business Media, LLC 2011

obtain for high-resolution NMR solution techniques and other proteins, such as membrane proteins, are not easily crystallized. Furthermore, structures obtained by crystallography do not always represent the biologically active conformation of the protein in solution. In these aforementioned situations, we can turn to complementary, albeit lower resolution approaches, which enable analysis of proteins under experimental conditions that closely resemble their native environment and often at much lower sample concentrations required for crystallography or NMR. Described herein is one such alternative structural approach which is well suited to all biomacromolecules, including large dynamic protein complexes. The approach utilizes the specific covalent attachment of a paramagnetic species, in this case a pair of stable nitroxide spin labels, via mutagenically introduced cysteine residues on the protein. Electron Paramagnetic Resonance (EPR) spectroscopy is then used to extract distances in the range of 8–25 Å from the dipolar interaction between the magnetic moments of the two introduced EPR spin labels. Multiple measured distances can then be used as constraints to test or define folding patterns of the protein under investigation. The current most commonly used analogous approach for measuring distances in biomacromolecules is Fluorescence Resonance Energy Transfer spectroscopy (FRET). The main advantage of using EPR for measuring distances over the FRET method is that only a single reporter label is required and thus complex labeling strategies need not be devised to attach the label at specific locations. Further, in contrast to FRET, the EPR spin labels are relatively small, comparable in size to that of a tryptophan residue.

Most biological macromolecules are EPR silent with net zero electronic spin and thus require the introduction of a paramagnetic species. Hubbell and colleagues pioneered the use of Site-Directed Spin Labeling (SDSL) as a method to introduce unpaired electrons into a protein fold, enabling analysis via EPR spectroscopy (1–3). The technique involves using routine molecular biology methods to introduce a cysteine residue into the protein backbone onto which a sulfydryl specific nitroxide spin label is then covalently attached via the formation of a disulfide bond (Fig. 1). The specific placement of these labels, as described herein, then allows for the targeted analysis of protein structure or conformation.

Modern day advancements now allow for rapid mutagenesis reactions using very standard routine procedures. User friendly kits are available on the market enabling simple mutagenesis reactions for even the inexperienced scientist. As cysteine residues are targeted for labeling by the EPR spin label, the site directed mutagenesis approach first requires a functional "cys-less" construct as a template for cysteine introduction. It is also highly desired that the function and structure of the protein under investigation is not perturbed upon the removal of the native cysteines. Once the

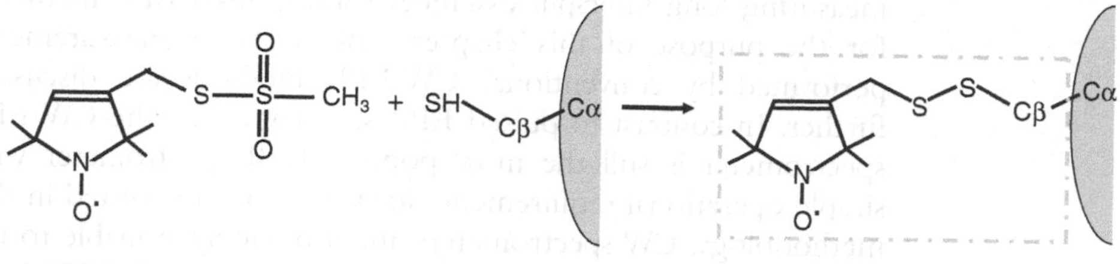

MTSSL Protein-cysteine Spin labeled side chain 'R1'

Fig. 1. Covalent attachment of the sulfydryl-specific nitroxide spin label (methanethiosulfonate, MTSSL) onto the backbone of a protein via formation of a disulfide bond with a cysteine residue. The EPR signal arises from the unpaired electron on the nitroxide group. The resulting spin-labeled side chain is generally referred to as R1.

cysteines are introduced at the target locations, the mutant protein can be recombinantly expressed in a bacterial *Escherichia coli* system, purified by chromatography methods and then modified with the EPR spin label.

Following spin labeling, several EPR measurements can be used to extract information regarding the protein structure. First, by measuring the EPR signal from a single attached spin label, structural information can be inferred from the mobility or solvent accessibility data for the spin label side chain (2, 4, 5). In conjunction with cysteine-scanning, the sequence dependence of these mobility and accessibility measurements can then be used for analysis of protein folding. Second, when pairs of spin labels are attached to the protein backbone, as is the focus of this chapter, the spin–spin interactions between the two nitroxides can result in broadening of the EPR spectrum. The dominant effect from the interspin interaction on the double-labeled EPR spectrum is the dipole–dipole interspin interaction. The distance information is extracted from the broadened EPR spectrum which can then be used to provide local short or long-range constraints for distinguishing between or building model structures. Typically, the dipole–dipole distance measurements which can be observed by continuous wave (CW) EPR spectroscopy are in the range of 8-25 Å. This distance range can further be extended to distances of up to approximately 80 Å using pulsed EPR methods (6). In many situations, the distance information is used in conjunction with the singly labeled EPR data from the mobility and accessibility measurements previously mentioned, allowing conclusions regarding the backbone fold to be drawn from both sets of data. However, as a stand-alone approach, the EPR distance method is also well suited for monitoring conformational changes in proteins in response to their activation by a physiological relevant trigger, such as ligand binding, membrane insertion or pH change. For further details on the SDSL-EPR method and structure determination, the reader is directed to many excellent reviews of this method (7–13). Despite the increase in popularity for measuring long interspin distances using pulsed EPR methods, for the purpose of this chapter, only distance measurements performed by conventional CW-EPR methods are discussed further. In contrast to pulsed EPR spectrometers, the CW-EPR spectrometer is still the most popular EPR spectrometer with simple operational requirements for the novice interested in this methodology. CW spectrometers are also widely available to the majority of researchers and the careful design of EPR probe pairs can address most structural questions.

CW-EPR methods have been used to obtain distance measurements in a multitude of experimental systems, including soluble proteins, membrane proteins (14), and nucleic acids (15). Secondary structure analysis (16), tertiary structure analysis (17), spatial mapping of quaternary structure (18), and conformational changes (19–21) are all achievable with these methods. However, due to the short range distance limit of ~25 Å, CW-EPR distance measurements are most effective for determining local protein folds. Thus, depending on the folding question being asked, a rational approach must be taken in the design of double cysteine mutant pairs that enables optimal testing or construction of a model by distance measurements. Further, obtaining a precise distance from the broadened EPR spectra of a double-labeled sample using CW-EPR is often complex without the detailed knowledge of local structure, side chain conformation and dynamics of the spin label (22, 23). An important emphasis of this chapter is to then perform the EPR distance measurements on multiple EPR spin labeled double mutants in order to rely on patterns of distances to deduce structural information. This is commonly required as a reliable conclusion is unlikely to be drawn from a single distance measurement.

In this chapter, the protocols for the site directed labeling of a protein, the collection of CW-EPR measurements at room temperature and the subsequent analysis of the CW-EPR spectra are described. Two semiquantitative analysis methods for extracting distances from CW-EPR spectra collected at room temperature are presented, each dependent on the molecular weight of the protein system under examination.

2. Materials

2.1. Purification and Labeling of Protein Samples with a Nitroxide Spin Label

1. Molecular biology materials (polymerase enzymes, plasmid extraction materials *etc*) for performing site-directed mutagenesis or a commercially available mutagenesis kit, such as the Quikchange II Site Directed Mutagenesis Kit (Stratagene, La Jolla, CA) or the Phusion™ Site Directed Mutagenesis Kit (Finnzymes Oy, Finland). Compatible competent *E. coli* cells for protein expression.

2. Protein sample of interest containing engineered cysteine residue(s), purified by standard procedures (see Note 1). Materials required for performing protein purification are user specific but may include culture media (Luria Bertani Broth), Isopropyl β-D-thiogalactoside (IPTG, Amresco, Solon, OH), antibiotics for growth selection, chromatography resins, and associated buffers. Purity of protein samples should be judged by sodium dodecyl sulfate-polyacrylamide gel electrophoresis (SDS-PAGE) (24). Purified and concentrated samples should be immediately labeled or stored at −70°C under reducing conditions in the presence of at least 1 mM dithiothreitol (DTT, Amresco).
3. Centriprep and/or Centricon filters (Millipore, Bedford, MA) with appropriate membrane molecular weight cut-off for concentration of protein samples. An alternative concentration method is with the use of Aquacide II (Calbiochem, La Jolla, CA).
4. EPR-labeling buffers. As no specific buffer is required for labeling, it is recommended that a buffer most appropriate for the proteins stability be used. Tris–HCl (Amresco), MOPS (Amresco), HEPES (Amresco), and phosphate buffers are all suitable for use within their appropriate pH range, noting that a higher pH value often results in a better labeling efficiency. Generally, a pH >7.2 is optimal for labeling. Reducing agents must not be present during (and after labeling) so as not to inhibit the labeling process. Salt (NaCl, KCl) should be included for protein stability where appropriate.
5. Dialysis tubing with an appropriate molecular weight cut-off for the protein under investigation (12–14 kDa, Sigma Aldrich, St. Louis, MO).
6. Gel filtration chromatography columns (recommended commercial columns include PD-10 or Sephadex 75 (GE Healthcare, Uppsala, Sweden)).
7. (1-oxyl-2,2,5,5-tetramethylpyrroline-3-methyl)methanethiosulfonate spin label (MTSSL, Toronto Research Chemicals, Toronto, Canada) is the most commonly used nitroxide spin label to produce the desired R1 side chain (Fig. 1), although other reagents can be used (see Note 2). Spin labels are most often supplied in a solid powder form and stocks prepared to a desired concentration of 0.1 M in 100% dimethyl formamide (DMF, Amresco) (see Note 3). This solution should be stored at −20°C and wrapped in foil to avoid exposure to ultraviolet (UV) light.
8. MTSSL spin label standards. A set of standards of an appropriate range that encompasses the labeled proteins theoretical spin concentration should be prepared by serial dilution with

distilled H_2O from the spin label stock (e.g. 5, 10, 25, 50, 75, 100, and 150 μM). TEMPOL (4-hydroxy tempo, Aldrich Chemical Company) can also be used as a standard.

2.2. EPR Spectroscopy

1. An X-band (microwave frequency ~9.5 GHz) CW-EPR spectrometer operated as per standard procedures with a microwave power supply (Klystron or Gunn diode) and rectangular waveguide, a variable field electromagnet with power supplies to generate and modulate a uniform magnetic field of several thousand Gauss, a high-Q resonant cavity (i.e., capable of storing thousands of times more energy than is dissipated on the walls), signal amplification circuitry, console and computer for data acquisition and analysis. Commercial spectrometers include the Bruker EMX, Bruker ElexSys E500 CW-series, or Varian E9 or 109 CW-series.
2. Spectra can be collected with a standard rectangular resonant cavity (TE_{102}, ER4119HS resonator (Bruker BioSpin, Billercia, MA)). Alternatively, the use of a loop-gap resonator or a dielectric resonator (Bruker ER4123D) can be used to increase the microwave magnetic field at the sample and can improve signal to noise with as little as 5 μL of sample.
3. Appropriate sample capillary tubes (quartz or suprasil) or gas-permeable TPX sample capillaries for a loop-gap resonator (Molecular Specialties Inc., Milwaukee, WI; Wilmad Lab Glass, Buena, NJ). For multiple samples, a cost effective option for single use sample tubes are 50 μL borosilicate glass capillary tubes (inner diameter = 0.5 mm, VWR Scientific, West Chester, PA; Drummond Scientific Company, Broomall, PA) loaded via capillary action and sealed at one end with Critoseal (Krackeler Scientific Inc., Albany, NY).
4. 4mm OD quartz tube (Wilmad Glass) for use with a rectangular cavity.
5. EPR is hundreds of times more sensitive than NMR and thus requires a much lower sample concentration. An ideal concentration range of 50–200 μM for the spin labeled sample is sufficient for collection of an optimal signal, but concentrations down to 5–10 μM can be used for spectrum collection. Sample requirements are thus user specific and can range from ~2–5 μL (10–500 μM spin label concentration) for a loop gap resonator to ~50–100 μL of 20–50 μM for a standard rectangular cavity. A consequence of a lower signal concentration is that a longer data acquisition time for increased signal averaging is required.
6. EPR analysis software. Most commercial software programs accompanying the spectrometer are capable of correcting spectra for baseline distortion or performing the double integration required for spin count analysis (see Subheading 3.3)

(Bruker BioSpin). These programs can also be used for measuring the spectral parameters (e.g., line heights, widths) required for distance analysis. Alternatively, programs such as LabVIEW (National Instruments, Austin, TX) can be applied to EPR spectra to extract required parameters.

3. Methods

3.1. Experimental Design: Proximity Mapping to Determine Protein Folds

SDSL-EPR is often used to complement existing NMR or crystallography data. In undefined regions, the design of appropriate distance pairs can reveal the relative arrangement of secondary structural elements or deduce protein tertiary folds from distance geometry mapping. The secondary structural elements in proteins can be investigated with distance measurements based on their unique periodicities of 3.6 and 2 for an α-helix and β-strand element, respectively. Creating an initial series of three double mutant pairs, such as (*i*, *i*+2), (*i*, *i*+3), and (*i*, *i*+4), where *i* represents the first-labeled cysteine/reference residue and the integer indicates how many residues away the second-labeled cysteine is positioned, can reveal secondary structure by observing a "pattern of proximities" (Fig. 2). Distances obtained from the initial set of measurements should then further guide the process of refinement of the structural model by the design of additional double mutant pairs. As the distances measured are between flexible nitroxides and not the protein backbone, multiple distance sets are necessary to reduce any potential bias due to the existence of multiple rotamers of the spin label side chain (25).

These short distances that can be measured by CW-EPR are not only limited to analysis of secondary structural elements, but can also provide geometric details for the assembly of the identified secondary structural elements or arrangement of other protein subunits in multicomponent systems. Tertiary and quaternary information is obtained from an initial set of distances, as described previously, prior to further refinement by additional distance pairs. An example of where these types of measurements could be suitable is provided in the (*i*, *i*+18) pairing of the β-turn in Fig. 2b. Another example is monitoring qualitative spectral trends or changes in the spin–spin interactions to understand the oligomeric nature and arrangement of a multicomponent system, such as an ion-channel (Fig. 3). Details of an inner helix pore of a channel can be revealed from the periodic pattern of the spin–spin interaction when cysteine scanning through consecutive residues is performed. The oligomeric assembly or the number of monomers forming the oligomer of the channel can be obtained by titration of the individually labeled subunit with an unlabeled subunit (26). The signal amplitude is enhanced as the concentration

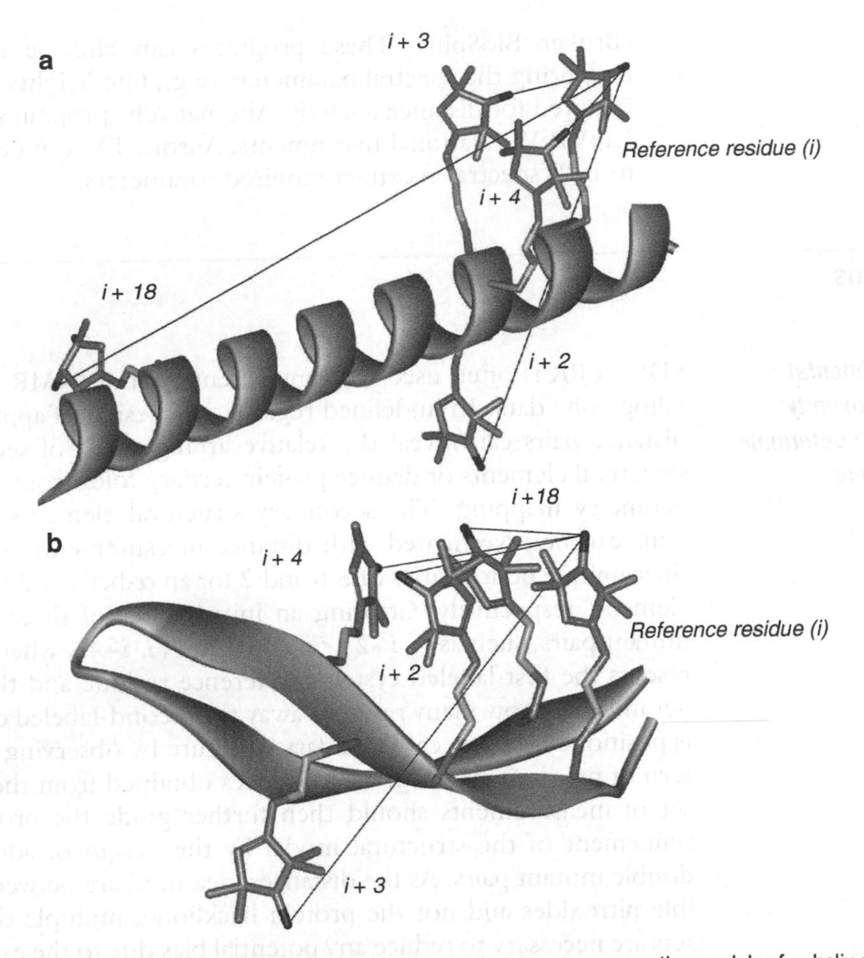

Fig. 2. Design of double spin-labeled mutant pairs to distinguish between two competing models of α-helix and β-strand secondary structural elements. The spin label R1 side chain is shown for illustration, noting that only one spin-label pair is present in any one experimental sample. Due to the unique periodicities of the two protein secondary structural elements, trends in distances measured by CW-EPR can be used for structural elucidation. (**a**) In the α-helix structure, two amino acid residues apart (*i*, *i*+2) will be on opposite sides of the helix while pairs three (*i*, *i*+3) or four (*i*, *i*+4) residues apart will be in close proxmity on the same side of the helix and result in a shorter measured interspin distance. The long distance pair (*i*, *i*+18) in this helical model would not be detectable using methods outlined in this chapter for CW-EPR. (**b**) In a competing model of β-strand elements, the trends in distances are almost reversed. Spin-labeled pairs two residues apart (*i*, *i*+2) will be positioned adjacent to each other while distance pairs (*i*, *i*+3) and (*i*, *i*+4) should result in longer measured interspin distances than those observed by the corresponding α-helix model in (**a**). An example measurement for determining the tertiary fold by CW-EPR is also shown for the spin-labeled pair (*i*, *i*+18) which are positioned in close proximity in the β-turn structural model.

of unlabeled monomers increases; with small oligomers displaying the greatest recovery of the signal from dilution with small concentrations of the unlabeled monomers. Further, the proposed structural response of a channel can be tested by measuring a single distance which is postulated to undergo an increase in its interspin distance or a decrease in spectral broadening, when transiting from a closed to open state. The methods that follow allow for the design and implementation of the above-mentioned

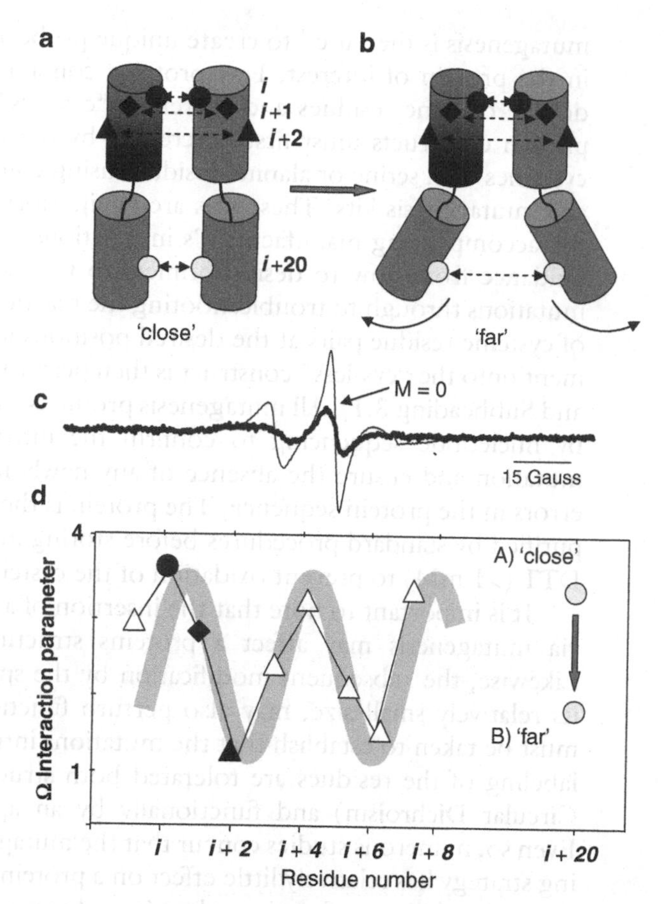

Fig. 3. Cartoon representation showing the use of the spin–spin interaction parameter Ω to elucidate the structure and conformational arrangement of two oligomeric subunits of an ion channel upon activation. (**a**) Each oligomeric subunit of the channel is labeled at consecutive residues numbered from residue "*i*" as represented by a filled circle, (*i*+1) by a filled diamond, and (*i*+2) by a filled triangle, etc. The spin-labeled channel is reconstituted into a membrane bilayer for each of the mutants and the spin–spin interaction parameter obtained from the ratio of the normalized amplitude of the central resonance line (*M*=0) between the single- and double-labeled spectra. (**b**) The activated open channel form results in the movement of the C-terminal domains away from each other. The movement from a short to long distance accompanying channel opening can be detected by the placement of a spin label at residue (*i*+20) on the inner-pore side, denoted by the open circle. (**c**) Corresponding EPR spectrum at room temperature of the closed channel form (*thick line*) and open channel form (*thin line*) for residue (*i*+20), normalized to the same number of spins. There is significant broadening and loss of intensity for the closed channel form reflecting a decrease in the distance between the two nitroxides. (**d**) Predicted Ω spin–spin interaction values for labeled residues, assuming complete uniform labeling of the channel was achieved for each mutant site. A strong spin–spin interaction is indicated by a high Ω value and vice-versa. The pattern observed indicates a helical structure.

experiments under physiological temperatures by simply measuring the spin–spin distance using a quantitative approach.

3.2. Labeling: Introduction of the Nitroxide Spin Label

To prepare the protein sample, it is essential to have an efficient plasmid for the overexpression of the desired protein construct. Generally, it is simplest and most cost effective to express in *E. coli*, but ideally any expression system that produces sufficient yields of the recombinant protein can be chosen. Routine site-directed mutagenesis is then used to create unique probe attachment sites in the protein of interest. Few proteins contain zero, single or double cysteine residues and therefore "cys-less" or "cys-light" protein constructs must first be created by replacing the native cysteines with serine or alanine residues using commercially available mutagenesis kits. These kits are straightforward to use and the accompanying manufacturer's instructions provide extensive guidance from how to design primers to introduce the desired mutations through to troubleshooting the reaction. Introduction of cysteine residue pairs at the desired positions for probe attachment onto the "cys-less" construct is then performed (see Note 4, and Subheading 3.1). All mutagenesis products should be verified by nucleotide sequencing to confirm the introduction of the mutation and ensure the absence of any newly introduced PCR errors in the protein sequence. The protein is then expressed and purified by standard procedures before storing in the presence of DTT (>1 mM) to prevent oxidation of the cysteine residues.

It is important to note that the insertion of a cysteine residue via mutagenesis may affect a proteins structure or function. Likewise, the subsequent modification by the spin label, despite its relatively small size, may also perturb function. Special care must be taken to establish that the mutations introduced and the labeling of the residues are tolerated both structurally (e.g., by Circular Dichroism) and functionally by an appropriate assay. Even so, numerous studies concur that the mutagenesis and labeling strategy has relatively little effect on a proteins backbone fold, thermal stability, or function when introducing cysteine residues at new positions or labeling with spin labels, particularly if surface residues are targeted (27–29).

A general method for achieving complete labeling of the protein is described, including a method for ensuring the effective reduction of cysteine residues prior to modification with the nitroxide spin label.

1. Purified protein (>5 nmol) is reduced in an appropriate labeling buffer by the addition of 1 M DTT to a final concentration of 5–10 mM. The sample is incubated at 4°C for 12 h to ensure complete reduction of cysteine residues.
2. Prior to the addition of the spin label, all DTT (as well as other reducing agents) must be effectively removed. Residual DTT has two effects on the spin label; it may inhibit conjugation of the disulfide bond formation between the cysteine and the spin label, and it may also reduce the unpaired electron spin signal. Removal of the DTT is achieved by gel filtration into the labeling buffer or by extensive dialysis (3×500 mL buffer changes over 24 h).
3. Immediately after the reducing agent has been removed, the nitroxide spin label (MTSSL stock) is slowly introduced dropwise to the protein solution, while vortexing, to generate the

side chain R1 (Fig. 1). A 5–10 fold molar excess of spin label over cysteine is added to ensure complete labeling. This amount is variable and is often dependent on the accessibility of the cysteine residues (see Note 5). The spin label should be dissolved in an organic solvent, such as DMF, as described in Subheading 2.1. Avoid adding greater than 1% (v/v) DMF to the protein solution. Thus, a suitable range of protein concentration for labeling is between 20 and 150 μM.

4. The labeling reaction is incubated for ~4 h at either 4°C or room temperature (based on the proteins stability) with gentle rocking. An optional addition of a second aliquot of spin label stock, up to 1–2 × molar excess is recommended prior to further incubation overnight at 4°C.
5. Unreacted spin label is removed at 4°C by either gel filtration chromatography in the desired buffer, or by extensive dialysis, or by other chromatography procedures (ion exchange, affinity, etc.) or by a combination of chromatography followed by dialysis. Once again, buffers used here must be free of reducing agents.
6. The labeled protein is concentrated using a centrifugal concentrator device with appropriate molecular weight membrane cut-off or Aquacide II to >10 μM protein concentration.
7. Protein concentration and thus cysteine concentration is determined by an appropriate assay: A_{280} for the protein (30) under investigation or by a standard Lowry (31), bicinchoninic acid (32), or Bradford (33) colorimetric protein assay using bovine serum albumin as a standard. The extent of labeling is determined by EPR spectroscopy as outlined in the following section.

3.3. EPR Spectroscopy: "Spin Count" to Confirm Labeling of Protein Sample

In a CW-EPR experiment, the applied magnetic field is swept while the frequency is held constant. For the EPR experiments described in this chapter, the operating frequency of the spectrometer is X-band (~9.5 GHz). The unpaired electron in the pπ orbital of the N–O bond, from the introduced nitroxide spin label, behaves as a magnetic dipole within the field which can then align itself parallel or antiparallel to the external field. These two orientations exist as states of low or high energy, respectively. When the energy levels between the two states exactly matches the energy of the microwave radiation ($\Delta E = h\nu$), transitions between the spin states occur, providing an absorption spectrum. Further, the unpaired electron also interacts with the nitrogen (^{14}N) nucleus further splitting the EPR signal. The integrated area of the EPR absorbance spectrum is proportional to the number of spins present in the sample. The first derivative of this absorption spectrum then provides the characteristic three peak EPR lineshapes used in the dipole–dipole distance analysis (Fig. 4). The central resonance peak corresponds to the single unpaired electron signal and the

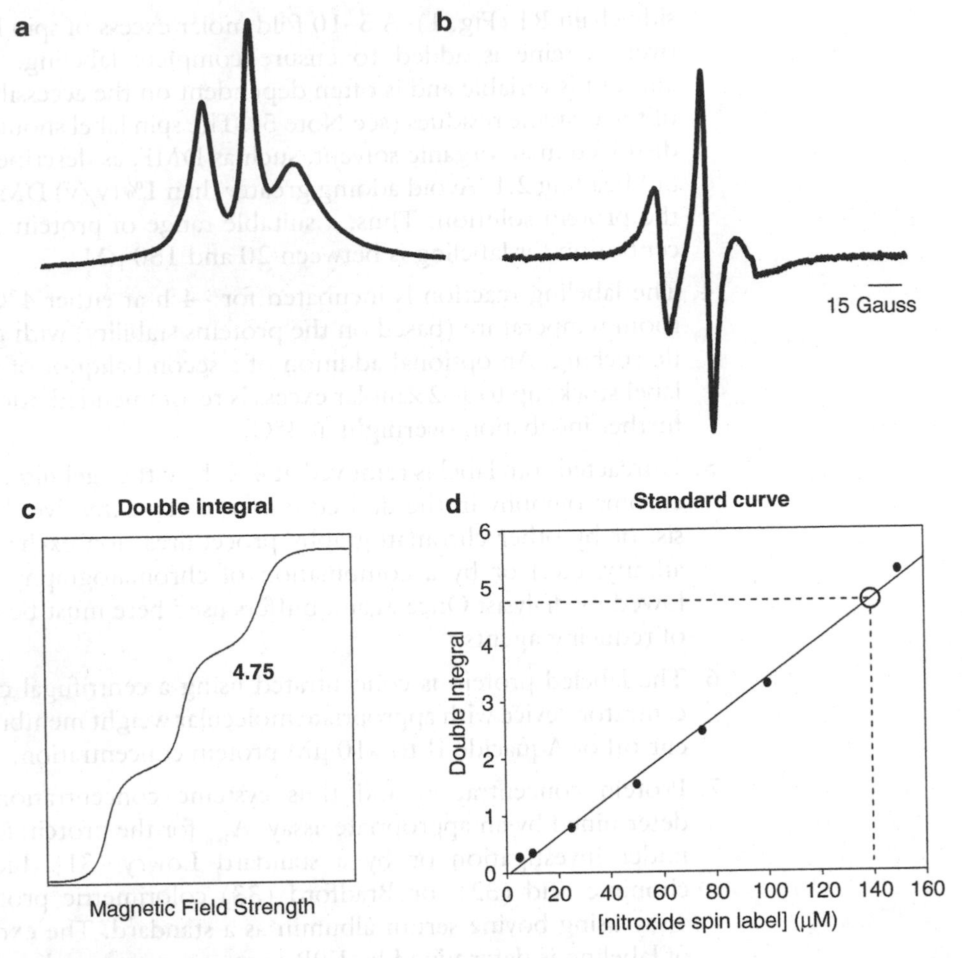

Fig. 4. The CW-EPR line shape and double integration of the first derivative CW-EPR spectrum to quantitate the concentration of spin label; "spin count." (**a**) An experimental nitroxide-labeled protein sample CW-EPR absorption spectral signal recorded at room temperature. The area beneath the EPR signal is proportional to the concentration of the EPR active species and the spectra are normalized to unit concentration by dividing the integral of the absorption spectrum. (**b**) Corresponding first derivative CW-EPR spectrum. The central peak results from the unpaired electron while the two hyperfine peaks result from electron–nuclei interactions. (**c**) Double integration of an experimentally labeled nitroxide sample. (**d**) Using double integral values, a standard curve is constructed to obtain the spin concentration ([spin]) of the labeled protein sample. The percentage labeling efficiency is obtained according to [spin]/[cysteine] × 100, where [cysteine] is measured from either absorbance or protein assay. For example, the measured double integral (4.75) for the above EPR spectrum was used in conjunction with the nitroxide standard curve to obtain a [spin] of ~140 μM. Thus, resulting labeling efficiency for a double spin-labeled protein sample with concentration of 74 μM ([cysteine] = 148 μM) was high at 95% (±10%).

high- and low-field hyperfine peaks are due to the described interaction of the unpaired electron with the nearby nitrogen nucleus.

The extent of labeling or spin labeling efficiency is therefore defined as the ratio of the nitroxide spin concentration, as obtained from the EPR absorbance spectrum, to the concentration of the available cysteine residues. The labeling ratio must be close to unity for optimal analysis of the interspin distance. The methods in the following section describe the measurement of the spin label concentration using double integration of the spectral line shape and comparison to a curve generated from the set of spin label standard samples.

1. Using the MTSSL stock, prepare a series of appropriate dilutions of spin label standards (50 μL each) as described in Subheading 2.1 and transfer to the EPR capillary. TEMPOL (4-hydroxy tempo) is a similar commercially available alternative that can also be used as a standard.
2. There is little variation in the operation of CW-EPR spectrometers; however, the manufacturer's instructions should be followed for optimal instrument operation. Likewise, many sample cavities are available to use as described in Subheading 2.2. For the purposes of this chapter, the methodology described is for a rectangular resonator (TE102 resonator) using a borosilicate glass capillary for sample loading.
3. Load 50 μL of the labeled protein sample into a borosilicate glass capillary tube (VWR) using capillary action and seal one end with Critoseal. Place the capillary within a clean 4 mm OD quartz tube and position in the middle of a standard rectangular EPR cavity, perpendicular to the static field, using appropriate collets.
4. Tune the cavity and collect the CW-EPR spectra at room temperature (22-24°C). It is recommended that the incident microwave power level should not exceed 5 mW, noting that the EPR signal intensity grows as the square root of the microwave power in the absence of saturation effects. Thus, care must be taken to avoid heat effects and saturation effects that can lead to broadening of the signal (see Note 6). For spin count analysis, a sweep-width of 140 G should be sufficient (see Note 7 if the double-labeled sample appears significantly broadened). To improve the signal-to-noise ratio, modify and optimize field parameters, including receiver gain (ensuring signals in spectrum are not clipped), time constant (an optimal time constant is achieved when the EPR line-width is 10 times greater than the length of the time constant), conversion time (increasing the conversion time helps resolve weak lines but results in longer sweep times), and field modulation amplitude (should be optimized to the natural line-width of the attached nitroxide (1.0–2.0 G for the nitroxide radical) at 100 kHz).
5. For generation of the nitroxide standard curve, spectral parameters must be kept consistent across all the samples. Likewise, for collection of the spectra of the spin labeled proteins, EPR parameters should also be kept consistent,

ensuring the microwave power is sufficiently below the saturation level.

6. To determine the labeling efficiency, the spin label concentration of the labeled protein sample is determined from the double-integration of the EPR spectrum. Most software accompanying the CW spectrometer (WIN-EPR) can perform simple double integration analysis of spectra. Generate a standard curve from the double integrated intensity values measured from the nitroxide standards versus the nitroxide standard concentration. The standard curve is then used in conjunction with the protein quantitation to obtain the spin count of labeled protein samples (see Note 8 and Fig. 4).

3.4. Continuous Wave Distance Measurements

When two spin labels are within 8–25 Å, the dipolar interaction between the two spins is reflected as a general broadening of the EPR spectrum and an associated decrease in the spectral amplitude of the double-labeled spectrum when compared to the composite spectrum of the two single noninteracting spins (Fig. 5). The extent of broadening is used to extract the distance information.

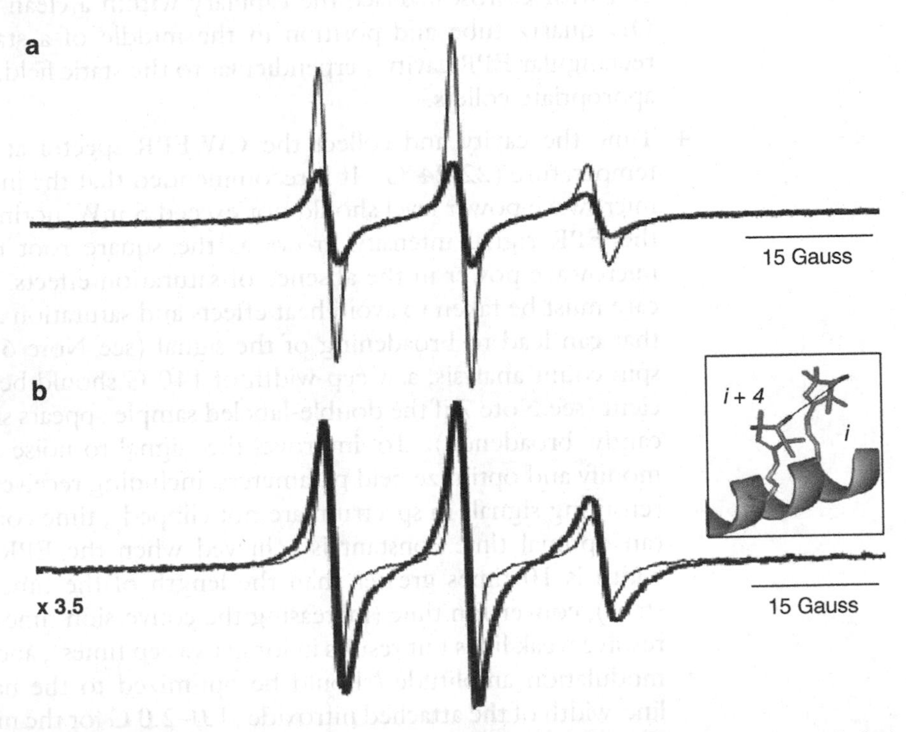

Fig. 5. EPR room temperature spectra of an (*i*, *i*+4) double mutant nitroxide pair positioned on a helical secondary structural region as shown in Fig. 2a. (**a**) The sum of the single noninteracting spectra (S - *thin line*) and the double-labeled interacting spectrum (D - *thick line*). (S) and (D) are normalized to the same number of spins. The decrease in peak intensities in the doubly labeled sample results from the broadening induced from the dipolar interaction between the two spins, corresponding to a distance of 14 Å (*unpublished, J Cooke*). (**b**) Double-labeled spectrum is scaled (~3.5 ×) for the comparison of amplitudes (and lineshape broadening).

In the rare case where the nitroxides adopt a specific, rigid geometry, both the relative distance and orientation between the two spin labels can be obtained from full spectral lineshape simulation (34). However, the majority of biological situations involving the introduction of paramagnetic species to measure distances include nitroxide spin labels that are not restricted in their motion and instead adopt a distribution of distances and relative orientations. Once again, complicated simulation analysis, such as convolution of the spectra from noninteracting spins with a broadening function, can be used to extract the distance information (35). Another quantitative spectral interpretation method for extracting the interspin distances from the dipolar broadening interaction is described by a Pake pattern-type broadening function (36). The ratio of the Fourier transforms of a doubly and a singly labeled spectrum gives a broadening function that is proportional to the R^{-3} distance between the spin labels. This Fourier deconvolution method has been experimentally verified for obtaining very accurate mean interspin distances in the range of 8–25 Å from the CW-EPR lineshape (16, 36, 37). In general, these quantitative analysis approaches are limited to experimental conditions where the EPR spin labels are immobilized on the EPR time scale or the so-called rigid lattice condition. This static case is most often achieved by collecting the EPR spectra under cryogenic conditions (~80 K). One major advantage of the SDSL-EPR approach is the ability to examine the sample under physiological conditions, including measurements at room temperature. Freezing the EPR samples to obtain rigid lattice conditions required for a more precise analysis of the interspin distance negates this advantage as well as inhibiting the real-time detection of conformational changes that may occur in solution at ambient temperatures. Further, freezing of samples can itself induce undesired protein conformational changes or bias the relative spin label rotamer populations (38).

To date, there is still an appropriate lack of a theoretical framework for performing an accurate and rigorous analysis of distances from spectral conditions, where the spin labels are mobile on the EPR time scale, as is often the case for samples collected at room temperature. Two semiquantitative methods can instead be applied, quite successfully.

The two cases for the analysis of distances by simple analysis of CW spectra collected at room temperature are described in the following section (see Note 9). The first analysis approach, based on the Redfield relaxation theory, is suitable for small proteins with a low-molecular mass of <20 kDa (Subheading 3.5) (39). Proteins of this size exhibit rapid tumbling in solution at room temperature of a few nanoseconds. Despite a warning on the qualitative nature of this approach, distances obtained using this analysis method have been shown to be very accurate when compared to the same distances obtained by X-ray crystallography (39). The second method described in this chapter is for labeled proteins at the other extreme of the size scale, with molecular weights >30 kDa. This protein class includes membrane proteins (Subheading 3.6) which exhibit rotational correlation times (τ_c) slow on the EPR time scale. If the two introduced nitroxide side chains happen to be rigidly attached to the protein, then the correlation time of the interspin vector is equal to the τ_c and the rigid lattice condition can be applied as the tumbling rate of the protein is slow relative to the dipolar coupling. However, most attached nitroxides have motion relative to the protein, and there is a lack of theory that can account for the dipolar interaction in conjunction with spin label motion. Nonetheless, it has also been shown that the internal motions of the nitroxide side chains have only a very small effect on the calculated interspin distances if the rigid lattice Fourier deconvolution method (36) is applied, especially if the protein molecular weight is >50 kDa (38). Analysis by the Fourier deconvolution method is not detailed further in this chapter, and the reader is directed to other reviews which further evaluate this method (9, 10, 40). Instead, an alternative simplified analysis method is described which measures a broadening parameter termed the "interaction parameter Ω" (29).

1. Calculate the estimated correlation time for the protein system under investigation. The relationship between the molecular weight of the protein and an estimated correlation time can be obtained from the Stokes–Einstein relationship given in Eq. 1:

$$\tau_c = \frac{M(\upsilon + h)\eta}{kT} \tag{1}$$

where M denotes the protein molecular weight, ν the partial specific viscosity (0.74 cm^3/g for proteins), h the hydration value (0.2 g water/gram protein), η the viscosity of solution, k the Boltzmann constant, and T the temperature. For proteins of a large molecular weight (>30 kDa), it is essential to ensure that the rotational correlation time is long enough to prevent averaging of the dipolar anisotropy. A value of >30 ns is sufficient and can be achieved by recording samples in the presence of 30–40% sucrose solution.
2. Collect CW-EPR spectra for the double labeled sample and each of the two noninteracting single-labeled spectra as described in Subheading 3.3. Normalize and average the two single noninteracting spectra to generate an EPR lineshape for the noninteracting spectrum (S) for comparison to the double-labeled interacting spectrum (D).
3. The EPR spectrum obtained for performing the spin count analysis for each sample in Subheading 3.3 may be sufficient

to use to extract the distance information. However, to improve signal to noise, especially for spin–spin interactions approaching the limit of the CW technique, optimization of spectral parameters is encouraged (see Note 10).

4. It is imperative that the EPR spectra are processed to remove baseline artifacts (offset and/or slope errors) prior to analysis. Increasing the scan-width up to 250 G can avoid the errors associated with base-line distortions of spectra.
5. Again, ensure that the EPR spectra are normalized for unit area by dividing by the double integral. That is, the single noninteracting spectrum generated from the average of the two single spectra (S) and the double spectrum (D) should represent the same number of spins.
6. Using the normalized spectrum (S) and (D), the distance between the paired nitroxides can be calculated from either of the following two methods depending on the molecular weight of the system under investigation: (1) spectral convolution for proteins of molecular weight <20 kDa or (2) measurement of the "Ω interaction parameter" for proteins of molecular weight >30 kDa.

3.5. Analysis of Distances: Spectral Convolution for Proteins <20 kDa

The spin–spin dipolar interaction is reflected both as a reduction in the EPR spectral amplitude and an associated broadening of the EPR line-width of a double-labeled sample with respect to the composite single-labeled spectrum. The spectral convolution scheme applied by Mchaourab and coworkers (39) uses lineshape broadening to obtain the spin–spin distances in solution at room temperature. The dominant mechanism is the modulation of the anisotropic dipole–dipole interaction by rapid rotational motion of the interspin vector. For this quantitative measure of the distance dependence from the interspin interaction, it is shown that relaxation effects lead to Lorentzian line broadening in which the ΔH (line-width at half height) is proportional to the spin–spin interaction R^{-6} (Fig. 6). The simple R^{-6} relationship between line-width and interspin distance is homogeneous and uniform across the spectrum and, provided the molecular weight of the protein system under examination is sufficiently low (*see below*), the dominant contribution to the overall broadening of the line-width from the modulation of the dipole–dipole interaction (ΔH_{dd}) is related to the correlation time according to Eq. 2:

$$\Delta H_{dd} = \frac{3}{10}\frac{\gamma^4}{g\beta}\left(\frac{h}{2\pi}\right)^3 \frac{\tau_c}{R_{dd}^6}\left(3+\frac{5}{1+v^2\tau_c^2}+\frac{2}{1+4v^2\tau_c^2}\right) \quad (2)$$

where γ is the gyromagnetic ratio of the electron, g is the electronic g factor, β is the Bohr magneton, and R_{dd} the interspin distance. It is important to note that as multiple distances are

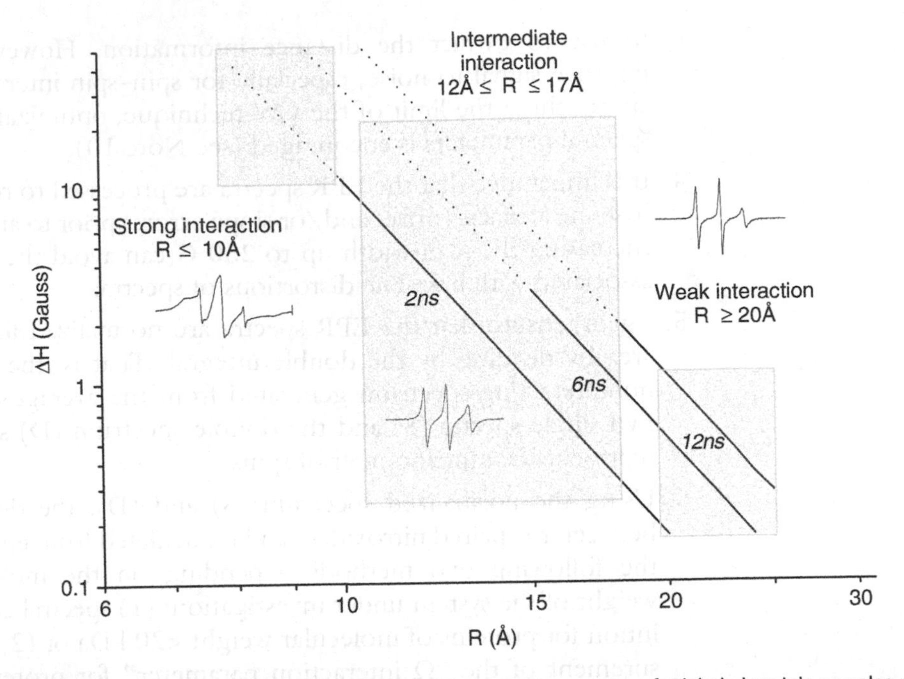

Fig. 6. Plot of expected ΔH values for interspin distances (R) at room temperature for labeled protein samples ranging in molecular weight from 5 kDa (~2 ns) to 30 kDa (~12 ns). Spectral broadening was calculated based on rotational correlation times of labeled protein sample at room temperature. The line-width is thus a function of both the interspin distance and the protein correlation time (τ_c). The plot can be used to estimate distances from experimental values of ΔH. Note, there are contributions to ΔH from unaveraged static dipolar interactions and, below 14 Å, contributions from through-space exchange interactions (*dotted line*). Predicted errors on the uncertainties of the measured interspin distances due to spin label position and orientation are estimated at approximately ±2 - 3 Å. A qualitative description can also be applied to spectra as follows: strong interaction ($R \leq 10$ Å), intermediate interaction ($12 \leq R \leq 17$ Å) and weak interaction (≥20 Å). Sample room temperature spectra of double spin-labeled mutants of cardiac muscle protein Troponin C are also included (*unpublished data*).

likely to exist in solution when the two spin labels are not fixed, the distance measured is heavily weighted in favor of the population of molecules with the shortest interspin distance as the line-width depends on R^{-6}.

As the magnetic dipole–dipole interactions can either be static or dynamic, this approach is only feasible for estimating the nitroxide interspin distance when the rotational correlation time of the protein is sufficiently short to average the static dipole interaction between nitroxides and thus best when the attached spin labels are undergoing rapid motion. Therefore, the practical limit for analysis using this method in aqueous solution is with a protein molecular of <20 kDa. This condition where the spin label mobility is high enough to modulate the anisotropy of the dipole–dipole interaction is determined by Eq 3:

$$\tau_c \leqslant \left(\frac{g\beta\Delta H_{dd}}{h}\right)^{-1} = \left(\frac{3\pi g^2\beta^2}{R_{dd}^3 h}\right)^{-1} \quad (3)$$

3.6. Analysis of Distances: Spin–Spin Interaction Parameter >30 kDa

For large molecular weight systems, the theory for interpreting spin–spin distances from CW spectra, where the label has arbitrary motion and the interspin vector is stationary on the EPR timescale is also lacking. One approach for extracting distance information is by monitoring the reduction in the spectral amplitude of the double-labeled interacting spectrum. The measured empirical parameter, "Ω," assigned the "spin–spin interaction" parameter by Perozo and coworkers, is defined as the ratio of the normalized amplitude of the central resonance line ($M=0$) between the single- and double-labeled spectra (29) (Fig. 3d). The Ω value is less quantitative than the convolution approach described for proteins <20 kDa, but it is more than ample for detecting spin–spin interactions when mapping geometric patterns relating to protein folds or distinguishing between two models by following trends in spin–spin interactions (Fig. 2). It is also a powerful approach for detecting spatial proximity accompanying protein function by measuring changes in Ω, as demonstrated in channel opening and closing events (14, 29, 41) (see Note 11).

As the signal amplitude for the double-labeled sample is inversely proportional to the strength of the spin–spin interaction, the value of Ω may approach values of ~3–4 if two labels are in close proximity and decreases to unity when no dipolar interaction is detected. Consequently, only broad descriptive categories should be applied to describe single distances measured by this method: close (<10 Å) (see Note 12), intermediate (10–15 Å), and for values of Ω ~1, far (>15 Å) (2). Once again, it must be emphasized that the strength of this approach is the ability to detect short spin–spin interactions under physiological conditions.

Once proximities between residues have been identified using either of these two analysis methods, and it is known that the sample can withstand nonphysiological temperatures, further well-established and experimentally verified quantitative approaches for extracting accurate distance information from the sample can be explored. These methods include those briefly mentioned in this chapter: computer lineshape simulation (13, 34), Fourier deconvolution methods (36), and pulsed EPR methods, such as DEER (6). The reader is also directed to an experimental evaluation of several of these methods for which the same distances measured by several approaches were found to be in reasonable agreement (42).

4. Notes

1. Ideally, the protein should be highly purified, although purity levels required for NMR and X-ray crystallography are not essential. Most contaminants are EPR "silent" but should be verified by collecting EPR spectra on the purified unlabeled sample or on a corresponding "cysless" sample.

2. The most commonly used spin label for covalent modification of cysteine residues is MTSSL although others, such as maleimide spin label (MSL, Toronto Research Chemicals) and iodoacetamide spin label (IASL, Toronto Research Chemicals) can also be used. Labeling efficiencies cannot be predicted and some spin labels may be more amenable to some proteins and labeling sites over others. We have found that MSL has an enhanced stability over MTSSL for samples which are exposed to room temperature conditions for extended periods of time.
3. Other solvents, such as acetonitrile or ethanol, can be used for preparing spin label stock solution. Ensure that the spin label stock solution is of a sufficient high concentration so as not to exceed 1% (v/v) protein solution when labeling.
4. In the design of the cysteine mutants, choose sites that are likely to be solvent exposed to improve labeling and thus the accuracy of the subsequent distance analysis. Modification of a surface exposed site is also most likely to result in minimal perturbation of a proteins structure.
5. An Ellman assay can be used to confirm the number of accessible sulfhydryl groups for labeling (43).
6. To check for saturation, acquire a spectrum and note the intensity. Then, decrease the microwave attenuation by 6 dB for which the microwave power has increased by a factor of four. Acquire another spectrum. If the signal intensity has increased by a factor of two, then no saturation has occurred. Continue to increase the microwave power to improve the signal-to-noise ratio until the onset of saturation is noticed. Use the largest microwave power for which no saturation was observed.
7. For double-labeled samples which are severely broadened and therefore may display low signal to noise, release of the attached spin label can be performed to accurately determine the labeling ratio. The attached spin label can be released by incubating the protein sample with 50 mM tris-(2-carboxyethyl)phosphine, (TCEP, Molecular Probes, Eugene, OR) and the integrated area used to obtain the spin concentration from a calibration curve prepared from standards freshly prepared in the same concentration of TCEP. Although TCEP is a reducing agent, the unpaired electron on the nitroxide spin label is stable under these conditions for periods of up to 1 h (44).
8. For distance analysis methods described in this chapter, it is necessary to achieve complete modification of the double cysteine protein sample with the nitroxide spin label. Thus, labeling conditions should be sought where maximum modification of both cysteine residues is achieved. Incomplete labeling could alter results as bias may be given to one of the cysteines and without

complete dipolar interaction, the double-labeled interacting spectrum (D) reflects a more mobile species, thereby influencing the lineshape and hence the distance analysis. Depending on the stability of proteins, increasing the pH of the labeling buffer may improve labeling ratios. Performing the labeling reaction at room temperature, as well as increasing the amount of excess spin label added may also improve labeling. In general, a surface exposed residue should label within minutes while a buried residue may require 4–8 h to label with a significant excess of spin label. If the protein to be labeled readily refolds, labeling may also be achieved in the presence of 6 M urea so that all residues are in an exposed environment. After labeling, the protein sample would require the gradual removal of the denaturant to refold to the native state.

9. Comparison of experimentally determined distances with predictions should be interpreted with care as the distances measured are between the nitroxide side chains and not between protein backbone atoms. Unknown nitroxide rotamer distributions from a label which itself can be up to 12 Å in length, can contribute a broad width of distance distributions, up to values of 10 Å. The distance estimated is the poorest for interspin distances less than 15 Å, where the overall mean error is ~6 Å (23). Modeling of both probe mobility and conformation using molecular dynamics methods have significantly improved agreement with experimental spin–spin distances for the interpretation of a $C_{\beta}-C_{\beta}$ distance with standard deviations expected to be ±3 Å (23, 45).
10. In practice, the limit for analysis by the CW method is ~20 Å. The dipolar interaction is often very difficult to detect when measuring longer distances in the range of 20–25 Å as there is a gradual loss in detectable line-broadening. Therefore, optimization of experimental conditions to obtain "clean" spectra with good baseline and high signal to noise are required to reliably extract distance information from the broadened spectra at the limit of the CW method with confidence. For distances between 20 and 25 Å, pulsed EPR methods, such as Double Electron-Electron Resonance (DEER), can be performed to confirm distances obtained by CW (6). The high sensitivity of the pulsed DEER method also allows for the extraction of multiple distances and distance distributions. However, DEER requires measurements to be performed under cryogenic conditions.
11. When using distance changes to monitor conformational changes, it should be made clear that the conformation of the protein backbone itself may cause changes in the spin label side chain flexibility and thus rotamer populations. It is possible that a decrease in mobility of one or both of the labeled

side chains can also contribute to the broadening effect of the double-labeled spectrum. Therefore, the mobility spectra of each single-labeled site should be observed in both functional states for the presence of a buried conformation prior to the analysis of the double-labeled spectrum. However, in practice, most conformational changes involve rigid body motions, and thus the environment of the spin label side chain located within well-defined segments of regular secondary structure should not significantly change.

12. Spin exchange (*J*) is a second type of interaction that can occur over very short distances when two probes are in sufficient proximity such that there is overlap of the orbitals of the unpaired electrons (<8 Å). The effects on the double-labeled spectrum are that the classical three-line EPR spectrum becomes reduced to a single feature. However, these interactions are rare in proteins when two residues are targeted with nitroxide spin labels. One recent example of spatial mapping using spin exchange interactions was shown by fibril formation in the α-synuclein system (18).

Acknowledgments

This work was supported by the Australian Research Council (ARC). LB is a recipient of an ARC APD fellowship and JC is a recipient of a Macquarie University Research Areas and Centres of Excellence Award (RAACE). The authors thank Mr Michael Howell for his editorial assistance.

References

1. Hubbell, W. L., and Altenbach, C. (1994) Investigation of structure and dynamics in membrane proteins using site-directed spin labeling. *Curr Opin Struct Biol.* **4**, 566–573.
2. Hubbell, W. L., Gross, A., Langen, R., and Lietzow, M. A. (1998) Recent advances in site-directed spin labeling of proteins. *Curr Opin Struct Biol.* **8**, 649–656.
3. Hubbell, W. L., McHaourab, H. S., Altenbach, C., and Lietzow, M. A. (1996) Watching proteins move using site-directed spin labeling. *Structure.* **4**, 779–783.
4. Klug, C. S., Su, W., and Feix, J. B. (1997) Mapping of the residues involved in a proposed beta-strand located in the ferric enterobactin receptor FepA using site-directed spin-labeling. *Biochemistry.* **36**, 13027–13033.
5. Columbus, L., and Hubbell, W. L. (2002) A new spin on protein dynamics. *Trends Biochem Sci.* **27**, 288–295.
6. Fajer, P., Brown, L., and Song, L. (2007) Practical Pulsed Dipolar ESR (DEER), in *Biological Magnetic Resonance* (Hemminga, M., and Berliner, L., Eds.), Vol. 27, Springer, New York, pp. 95–128.
7. Czogalla, A., Pieciul, A., Jezierski, A., and Sikorski, A. F. (2007) Attaching a spin to a protein - site-directed spin labeling in structural biology. *Acta Biochim Pol.* **54**, 235–244.
8. Schiemann, O., and Prisner, T. F. (2007) Long-range distance determinations in biomacromolecules by EPR spectroscopy. *Q Rev Biophys.* **40**, 1–53.
9. Fanucci, G. E., and Cafiso, D. S. (2006) Recent advances and applications of site-directed spin labeling. *Curr Opin Struct Biol.* **16**, 644–653.
10. Steinhoff, H. J. (2004) Inter- and intramolecular distances determined by EPR spectroscopy and site-directed spin labeling reveal

protein-protein and protein-oligonucleotide interaction. *Biol Chem.* **385**, 913–920.

11. Borbat, P. P., Costa-Filho, A. J., Earle, K. A., Moscicki, J. K., and Freed, J. H. (2001) Electron spin resonance in studies of membranes and proteins. *Science.* **291**, 266–269.
12. Hubbell, W. L., Cafiso, D. S., and Altenbach, C. (2000) Identifying conformational changes with site-directed spin labeling. *Nat Struct Biol.* **7**, 735–739.
13. Hustedt, E. J., and Beth, A. H. (1999) Nitroxide spin–spin interactions: applications to protein structure and dynamics. *Annu Rev Biophys Biomol Struct.* **28**, 129–153.
14. Perozo, E., Cortes, D. M., and Cuello, L. G. (1999) Structural rearrangements underlying K+–channel activation gating. *Science.* **285**, 73–78.
15. Cai, Q., Kusnetzow, A. K., Hideg, K., Price, E. A., Haworth, I. S., and Qin, P. Z. (2007) Nanometer distance measurements in RNA using site-directed spin labeling. *Biophys J.* **93**, 2110–2117.
16. Brown, L. J., Sale, K. L., Hills, R., Rouviere, C., Song, L., Zhang, X., and Fajer, P. G. (2002) Structure of the inhibitory region of troponin by site directed spin labeling electron paramagnetic resonance. *Proc Natl Acad Sci U S A.* **99**, 12765–12770.
17. Xiao, W., Poirier, M. A., Bennett, M. K., and Shin, Y. K. (2001) The neuronal t-SNARE complex is a parallel four-helix bundle. *Nat Struct Biol.* **8**, 308–311.
18. Chen, M., Margittai, M., Chen, J., and Langen, R. (2007) Investigation of alpha-synuclein fibril structure by site-directed spin labeling. *J Biol Chem.* **282**, 24970–24979.
19. Hanson, S. M., Francis, D. J., Vishnivetskiy, S. A., Klug, C. S., and Gurevich, V. V. (2006) Visual arrestin binding to microtubules involves a distinct conformational change. *J Biol Chem.* **281**, 9765–9772.
20. Radzwill, N., Gerwert, K., and Steinhoff, H. J. (2001) Time-resolved detection of transient movement of helices F and G in doubly spin-labeled bacteriorhodopsin. *Biophys J.* **80**, 2856–2866.
21. Xiao, W., Brown, L. S., Needleman, R., Lanyi, J. K., and Shin, Y. K. (2000) Light-induced rotation of a transmembrane alpha-helix in bacteriorhodopsin. *J Mol Biol.* **304**, 715–721.
22. Hustedt, E. J., Stein, R. A., Sethaphong, L., Brandon, S., Zhou, Z., and Desensi, S. C. (2006) Dipolar coupling between nitroxide spin labels: the development and application of a tether-in-a-cone model. *Biophys J.* **90**, 340–356.
23. Sale, K., Song, L., Liu, Y. S., Perozo, E., and Fajer, P. (2005) Explicit Treatment of Spin Labels in Modeling of Distance Constraints from Dipolar EPR and DEER. *J Am Chem Soc.* **127**, 9334–9335.
24. Walker, J. M. (2002) SDS Polyacrylamide Gel Electrophoresis of Proteins, in *The Protein Protocols Handbook* (Walker, J. M., Ed.), Humana, Totowa, NJ, pp. 61–67.
25. Altenbach, C., Cai, K., Klein-Seetharaman, J., Khorana, H. G., and Hubbell, W. L. (2001) Structure and function in rhodopsin: mapping light-dependent changes in distance between residue 65 in helix TM1 and residues in the sequence 306–319 at the cytoplasmic end of helix TM7 and in helix H8. *Biochemistry.* **40**, 15483–15492.
26. Langen, R., Isas, J. M., Luecke, H., Haigler, H. T., and Hubbell, W. L. (1998) Membrane-mediated assembly of annexins studied by site-directed spin labeling. *J Biol Chem.* **273**, 22453–22457.
27. Langen, R., Oh, K. J., Cascio, D., and Hubbell, W. L. (2000) Crystal structures of spin labeled T4 lysozyme mutants: implications for the interpretation of EPR spectra in terms of structure. *Biochemistry.* **39**, 8396–8405.
28. McHaourab, H. S., Lietzow, M. A., Hideg, K., and Hubbell, W. L. (1996) Motion of spin-labeled side chains in T4 lysozyme. Correlation with protein structure and dynamics. *Biochemistry.* **35**, 7692–7704.
29. Perozo, E., Cortes, D. M., and Cuello, L. G. (1998) Three-dimensional architecture and gating mechanism of a K+ channel studied by EPR spectroscopy. *Nat Struct Biol.* **5**, 459–469.
30. Aitken, A., and Learmonth, M. P. (2002) Protein Determination by UV Absorption, in *The Protein Protocols Handbook* (Walker, J. M., Ed.), Humana, Totowa, NJ, pp. 3–6.
31. Waterborg, J. H. (2002) The Lowry Method for Protein Quantitation, in *The Protein Protocols Handbook* (Walker, J. M., Ed.), Humana, Totowa, NJ, pp. 7–9.
32. Walker, J. M. (2002) The Bicinchoninic Acid (BCA) Assay for Protein Quantitation, in *The Protein Protocols Handbook* (Walker, J. M., Ed.), Humana, Totowa, NJ, pp. 11–14.
33. Kruger, N. J. (2002) The Bradford Method for Protein Quantitation, in *The Protein Protocols Handbook* (Walker, J. M., Ed.), Humana, Totowa, NJ, pp. 15–21.
34. Hustedt, E. J., Smirnov, A. I., Laub, C. F., Cobb, C. E., and Beth, A. H. (1997) Molecular distances from dipolar coupled spin-labels: the global analysis of multifrequency continuous wave electron paramagnetic resonance data. *Biophys J.* **72**, 1861–1877.
35. Steinhoff, H. J., Radzwill, N., Thevis, W., Lenz, V., Brandenburg, D., Antson, A., Dodson, G., and Wollmer, A. (1997) Determination of interspin distances between

spin labels attached to insulin: comparison of electron paramagnetic resonance data with the X-ray structure. *Biophys J.* **73**, 3287–3298.

36. Rabenstein, M. D., and Shin, Y. K. (1995) Determination of the distance between two spin labels attached to a macromolecule. *Proc Natl Acad Sci U S A.* **92**, 8239–8243.
37. Zhang, F., Chen, Y., Kweon, D. H., Kim, C. S., and Shin, Y. K. (2002) The four-helix bundle of the neuronal target membrane SNARE complex is neither disordered in the middle nor uncoiled at the C-terminal region. *J Biol Chem.* **277**, 24294–24298.
38. Altenbach, C., Oh, K. J., Trabanino, R. J., Hideg, K., and Hubbell, W. L. (2001) Estimation of inter-residue distances in spin labeled proteins at physiological temperatures: experimental strategies and practical limitations. *Biochemistry.* **40**, 15471–15482.
39. McHaourab, H. S., Oh, K. J., Fang, C. J., and Hubbell, W. L. (1997) Conformation of T4 lysozyme in solution. Hinge-bending motion and the substrate-induced conformational transition studied by site-directed spin labeling. *Biochemistry.* **36**, 307–316.
40. Xiao, W., and Shin, Y.-K. (2000) EPR Spectroscopic Ruler: The Deconvolution Method and its Applications, in *Biological Magnetic Resonance: Distance Measurements in Biological Systems by EPR* (Berliner, L. J., Eaton, S. S., and Eaton, G. R., Eds.), Vol. 19, Kluwer Academic/Plenum Publishers, New York, pp. 249–276.
41. Perozo, E., Cortes, D. M., Sompornpisut, P., Kloda, A., and Martinac, B. (2002) Open channel structure of MscL and the gating mechanism of mechanosensitive channels. *Nature.* **418**, 942–948.
42. Persson, M., Harbridge, J. R., Hammarstrom, P., Mitri, R., Martensson, L. G., Carlsson, U., Eaton, G. R., and Eaton, S. S. (2001) Comparison of electron paramagnetic resonance methods to determine distances between spin labels on human carbonic anhydrase II. *Biophys J.* **80**, 2886–2897.
43. Riddles, P. W., Blakeley, R. L., and Zerner, B. (1983) Reassessment of Ellman's reagent. *Methods Enzymol.* **91**, 49–60.
44. McHaourab, H. S., Berengian, A. R., and Koteiche, H. A. (1997) Site-directed spin-labeling study of the structure and subunit interactions along a conserved sequence in the alpha-crystallin domain of heat-shock protein 27. Evidence of a conserved subunit interface. *Biochemistry.* **36**, 14627–14634.
45. Fajer, P. G. (2005) Site directed spin labelling and pulsed dipolar electron paramagnetic resonance (double electron-electron resonance) of force activation in muscle. *J Phys Condens Matter.* **17**, S1459–S1469.

Chapter 7

Solution-State Nuclear Magnetic Resonance Spectroscopy and Protein Folding

Lisa D. Cabrita, Christopher A. Waudby, Christopher M. Dobson, and John Christodoulou

Abstract

A protein undergoes a variety of structural changes during its folding and misfolding and a knowledge of its behaviour is key to understanding the molecular details of these events. Solution-state NMR spectroscopy is unique in that it can provide both structural and dynamical information at both high-resolution and at a residue-specific level, and is particularly useful in the study of dynamic systems. In this chapter, we describe NMR strategies and how they are applied in the study of protein folding and misfolding.

Key words: NMR spectroscopy, Protein folding, Isotopic labelling, H/D exchange, HSQC, Diffusion

1. Introduction

The use of solution-state Nuclear Magnetic Resonance (NMR) spectroscopy is a powerful technique that is able to report at a residue-specific level, on both protein structure and dynamics – two important facets of conformational change during protein folding and misfolding. As NMR spectroscopy is a wide-ranging subject with many prominent workers in the field, it is beyond the scope of this chapter to cover all of the available methodologies; this chapter thus emphasises the most common NMR techniques that are applied to probe structure and dynamics in the context of the conformational states that a protein can adopt during folding and misfolding. This chapter focuses on three fundamental areas in NMR: sample preparation, 1D and 2D data acquisition/analysis, and discussion on the applicability of more advanced strategies to describe protein folding and misfolding.

Andrew F. Hill et al. (eds.), *Protein Folding, Misfolding, and Disease: Methods and Protocols*, Methods in Molecular Biology, vol. 752, DOI 10.1007/978-1-60327-223-0_7,

2. Materials

1. 10×M9 Salts: (67.9 g/L Na_2HPO_4, 30 g/L KH_2PO_4, 5 g/L NaCl).
2. Isotopes: ^{15}N labelling, 1 g/L ^{15}N ammonium chloride (Sigma). Adjust the pH to 7.2 with NaOH. ^{13}C labelling, 2 g/L ^{13}C-D-glucose (Sigma).
3. Vitamin supplement solution: Suggestions – Vitamins with amino acids or ammonium sulphate (e.g. Yeast Nitrogen Base, DIFCO); Sigma BME vitamins (100×). Suggested usage is 0.8 g/L Yeast Nitrogen Base without amino acids or ammonium sulphate (Difco) ×1 Sigma BME vitamins.
4. Labelling medium: (1×M9 salts, 100 μM $CaCl_2$, 2 mM $MgSO_4$, 1 g/L ^{15}N ammonium chloride, 4 g/L ^{12}C glucose, vitamin supplement solution, sterile water). For $^{13}C/^{15}N$ labelling, ^{13}C-D-glucose can be used at 2 g/L.
5. LB medium (16 g Tryptone, 10 g Yeast, 5 g NaCl).
6. Isopropyl β-D-1-thiogalactopyranoside (IPTG), (Sigma).
7. 0.1% (w/v) 4,4-dimethyl-4-silapentane-1-sulfonic acid (DSS) (Sigma).
8. 99% Deuterium Oxide (with or without 1% (w/v) DSS (^2H_2O or D_2O) (Sigma).
9. 99.9% Deuterium Oxide (^2H_2O or D_2O) (Cambridge Isotope Laboratories).
10. 1-oxyl-2,2,5,5-tetramethyl-3-pyrroline-3-methyl methane-thiosulfonate (MTSL) (Toronto Research Chemicals Inc).
11. HiTrap Desalting column (GE Healthcare).
12. Acetonitrile (Sigma).
13. 5 mM OD Precision borosilicate glass tubes (for use with a 5 mm probe) (Wilmad glass company).
14. A 500 MHz, 600 or 700 MHz spectrometer equipped with 4 channels and deuterium decoupling and a 5 mm ^1H,^{13}C,^{15}N room temperature or cryo probe (e.g. Bruker Avance III).

3. Methods

3.1. Preparing Isotopically Labelled Protein for NMR Spectroscopy

The most abundant isotopes of carbon and nitrogen, ^{12}C and ^{14}N, are not amenable to observation by solution-state NMR. Therefore, in all but a small number of experiments to be suitable for NMR detection, a protein must be isotopically labelled, often uniformly with ^{15}N (single-labelled) and/or ^{13}C (double-labelled). Deuteration or perdeuteration (triple-labelled $^{2}H/^{15}N/^{13}C$) may be used to reduce spectral complexity and improve relaxation properties, of particular importance for large (> ca. 25 kDa) molecules. In addition, the selective labelling of amino acids is not only useful for assignment and structure determination, but also provide a means of simplifying the spectra of very large protein systems.

3.1.1. Uniform $^{15}N/^{13}C$ Isotopic Labelling

1. Prepare the medium. Flasks containing M9 salts and water may be autoclaved; other components should be filter sterilised and added afterwards.
2. Take a single colony from a plate and prepare an overnight culture in LB medium (with appropriate antibiotics) at 37°C. Prepare 10 mL for every litre of culture to be inoculated.
3. Pellet the cells, resuspend in 1 × M9 and pellet the cells again (wash step).
4. Use these cells to inoculate 1 L of labelling medium.
5. Grow at 37°C, 200–250 rpm until OD600 ~0.5–0.6.
6. Induce expression with 1 mM IPTG for 3–5 h.
7. Harvest the cells by centrifugation and resuspend the cells in an appropriate buffer.

3.2. Deuteration/Perdeuteration

Large proteins (> ca. 25 kDa) have slow rotational ("tumbling") properties which results in faster transverse relaxation and an overall loss in sensitivity. This is because the desired magnetisation transfer that occurs through the protein's backbone during the pulse sequence is lost through various relaxation pathways, dominated to a large extent by sidechain protons (1, 2) and this compromises the detectable signal. The result is broad, overlapping resonances which complicates the analysis of spectra (if they are not broadened beyond the limit of detection). Perdeuteration (complete deuteration) substitutes deuterons for protons (^{1}H to ^{2}H) within each residue, which decreases the strength of the dipolar coupling interaction due to the smaller magnetic moment of ^{2}H. The reduced strength of the dipolar coupling interaction leads to slower transverse relaxation, which results in sharper peaks (3–5). Fractional deuteration (50–85%) can also significantly improve the resolution of ^{1}H in some cases (6, 7). In combination with experiments utilising a TROSY-based strategy, this has been useful for the characterisation of large protein systems (8–10).

For deuteration alone, uniform labelling can proceed as described above; however, all components (M9 salts, glucose, NH_4Cl, $MgSO_4$, $CaCl_2$) must be made in $^{2}H_2O$ (99.9% $^{2}H_2O$). This results in at least 80% deuteration and variations to produce fractional deuteration can also be performed. *Escherichia coli* also requires adaptation into a deuterated media for successful expression.

3.2.1. Preparation for Labelling Strategies Requiring Deuteration

1. Sterilise a 2 L Erlenmeyer flask and allow to completely dry either in an oven or at room temperature.
2. Prepare all components in 99.9% ^2H_2O (or in fractional amounts if required) and sterilise by filtration.
3. From a fresh plate, inoculate 5 mL of LB media (with antibiotics).
4. Incubate at 37°C, 200–250 rpm until slightly turbid (OD_{600} 0.2–0.4).
5. Remove a 1 mL sample and pellet the cells in a microfuge (2 min, 15,000×*g*). Resuspend the cells in 1 mL of M9 salts and pellet again. Remove the supernatant and resuspend the pellet in 10 mL of deuterated media (use ^{14}N and ^{12}C nitrogen and carbon sources, respectively).
6. Grow overnight (14–16 h) at 37°C.
7. Proceed with expression as described in Subheading 3.1.1.

Perdeuteration requires that all protons be exchanged for deuterons. Most deuterium is supplied as a 99% solution, and it is common practice to dissolve all media components to ensure a high level of deuteration, as described above. Some researchers have also used an approach, where all of the above solutions in H_2O, are freeze dried, and then resuspended in ^2H_2O to allow for proton exchange then followed by a second freeze dry step before resuspending in ^2H_2O. *E. coli* typically exhibits slower rates of growth in deuterated media and in the case of perdeuteration, may require more extensive adaptation, either by successive growth into increasing concentrations of ^2H_2O (80, 90, 95%, etc.) or may require several cycles of dilution and growth in deuterated media (11).

3.3. Selective Amino Acid Incorporation

Using a selective amino acid strategy largely depends on the information to be obtained and interpreted within a given protein sequence. More recently, there have been significant advances in selective labelling of methyl groups using precursors during growth, which have shown great promise in the study of larger protein systems; however, the methods associated with this technique is beyond the scope of chapter and is elegantly reviewed elsewhere (12). The production of proteins selectively labelled with amino acids can be hampered by isotopic dilution and the misincorporation of label at alternative sites as a result of transaminase (aminotransferases) activity which catalyses the transform of amino groups between α-amino and α-keto acids and which subsequently leads to nitrogen exchange (13). The rate at which this occurs is linked to the available pool of amino acids, the rates of synthesis and degradation of an amino acid, which is tied to the overall length of induction period during protein expression. The conversion of labelled amino acids to unlabelled amino acids through metabolic pathways can be overcome by adding an excess of the desired labelled amino acids together with an excess of appropriate unlabelled amino acids 20–30 min prior to protein induction to promote the inhibition of feedback mechanism. The length of time that allows for maximum protein expression, but minimises isotopic dilution needs to be assessed for a given protein. The use of auxotrophic cell lines is also a proven method for selective amino acid incorporation (14).

3.4. ^{1}H 1D NMR Provides a Fingerprint of a Protein's Conformation

Exploring the structure and dynamics of a protein and the conformational changes it undergoes generally requires a combination of both one-dimensional and two-dimensional NMR experiments. However, the most fundamental NMR experiment to probe a protein's structure is a proton (^{1}H) 1D experiment, and in its simplest form this can provide specific information regarding the integrity and conformation of a protein.

An advantage of the 1D is that it provides an overall fingerprint of the protein, and it does not require isotopic labelling. Each peak in the spectrum represents a proton and while it may be difficult to discern the peaks, there are features within the 1D itself which indicate whether indeed a protein is folded or not (Fig. 1a). Of interest is the amide (HN) region (6–11 ppm), within which overlaps the aromatic resonances (6–8 ppm). A wide dispersion of amide resonances is associated with folded protein. The aliphatic (HC) region (−1 to 6 ppm) is also of interest, particularly methyl groups, which fall between 1.5 and −1 ppm. Within a folded protein, the shielding that takes place near aromatic sidechains within the highly packed hydrophobic core leads to high-field shifted methyl groups (*ca.* 0.5 to −1 ppm). An examination of the Hα protons (4.4–6 ppm) can also report on the secondary structure content of a protein, with low-field peaks in the 5.5–6.25 ppm range corresponding to β strand conformations. Typically in a folded protein, one would expect to see a good dispersion of amide peaks between 6 and 11 ppm, the specific pattern of which is unique between proteins. In an unfolded protein, one would expect to see generally poorly dispersed amide, aromatic and alpha protons and a lack of high-field methyl groups. Intrinsically disordered proteins also show a narrow dispersion of amide resonances that lies within 8–8.5 ppm region similar to that seen within an unfolded protein.

Another important consideration with the 1D is the linewidth measurement, which represents the width of a given peak at half its height with the spectrum, as these are important indicators for both aggregation and dynamics. As a protein increases in size, its tumbling time increases and transverse (T_2) relaxation occurs much more quickly, resulting in a broadening of the linewidths.

3.4.1. Acquiring Proton (^{1}H) 1D Spectrum

1. Prepare a highly pure protein sample (>90% as judged by SDS-PAGE) in a suitable buffer (see Note 1) at a minimum concentration of 100 μM.

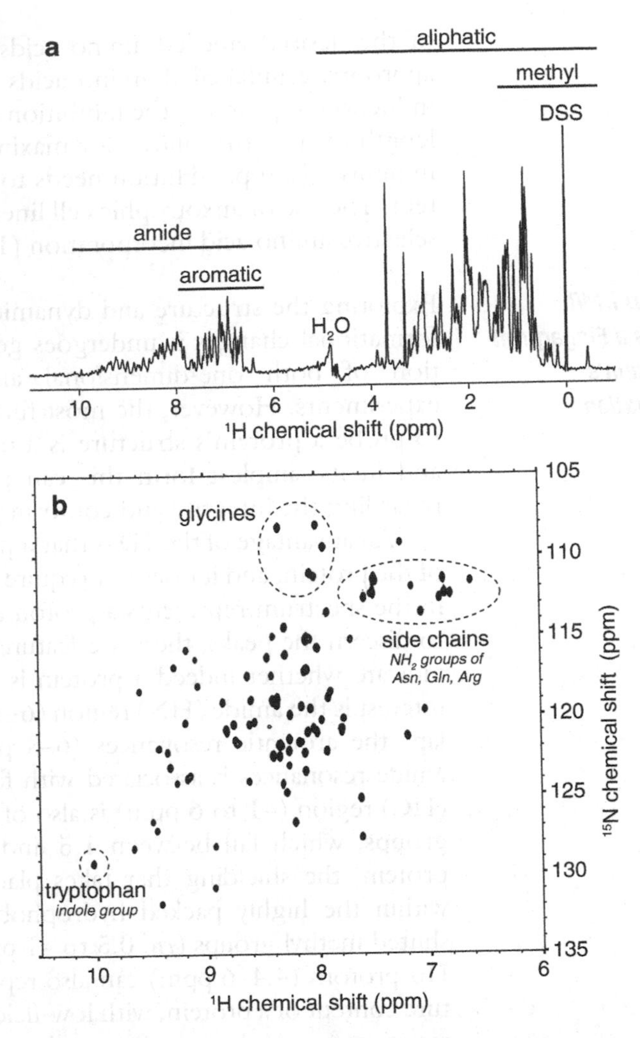

Fig. 1. A 1D proton spectrum represents the signature of a protein (**a**) Shown here is a spectrum of the PI3-SH3 domain recorded at 298 K. Several regions within a spectrum provide information about specific groups within a protein, in particularly the amide (HN) region (6–11 ppm) and the aromatics (6–8 ppm), where a wide dispersion of these resonances reflects a folded protein; similarly, the aliphatic (HC) region (−1 to 6 ppm) and the methyl groups (1.5 to −1 ppm) are also indicators of the degree of structure. (**b**) A ^{1}H-^{15}N HSQC spectrum of the same sample. The spectrum provides both structural and dynamical information about a protein through the analysis of both the chemical shift and the linewidth for each resonance, which corresponds to a given amino acid within a protein. Distinct resonances of particular amino acids groups (glycines, tryptophan, sidechain groups) can also be discerned. Using three-dimensional experiments, each resonance can be assigned to its corresponding amino acid residue.

2. To the sample add 99.9% D_2O to a final concentration of 5–10% (v/v). Optionally, a small quantity (0.1% w/v) of DSS may be added as a chemical shift reference (see Note 2).
3. Set and equilibrate the temperature, lock, tune and shim the spectrometer and use a simple proton experiment to calibrate the 90° pulse.

4. Run a presaturation experiment to find the exact frequency of the water resonance, which should be placed on-resonance (i.e. at the carrier frequency) to optimise water suppression (see Note 3).
5. Set up a one-dimensional pulse sequence using the calibrated 90° pulse. A spectral width of 14 ppm is common for proteins. A time domain of 2,048 points with 128 scans and a 1.5 s relaxation delay is a useful starting point for small proteins (see Note 4).
6. Examine the dispersion properties of the amide, aromatic and methyl regions within the spectrum as well as the linewidths of these peaks.

3.5. Two-Dimensional Correlation Spectroscopy to Probe Residue-Specific Events Through Chemical Shift Perturbations

At the heart of NMR spectroscopy is the chemical shift, which is a highly sensitive reporter of the local environment, and is influenced by both local and global events and factors, such as pH, temperature and protein concentration. While 1D spectra can provide some basic chemical shift information, the massive amount of overlap prevents, in general, the identifical of individual residues. Two-dimensional spectroscopy solves this problem by resolving resonances in a second chemical shift dimension. For example, in a ^{1}H-^{15}N heteronuclear single quantum coherence (HSQC) spectrum, every amide and sidechain NH in the protein gives rise to a crosspeak defined by the proton and nitrogen chemical shifts (Fig. 1b). Similarly, a ^{1}H-^{13}C HSQC experiment (or, more commonly, the qualitatively similar heteronuclear multiple quantum coherence (HMQC) experiment) can report on sidechain carbon resonances. As each crosspeak is unique to each residue (excluding overlap), it allows for local and global events to be probed with residue-specific resolution, when an assignment is available. The assignment of protein spectra has been reviewed elsewhere (e.g. (15)).

By examining the crosspeak dispersion properties of the spectrum, one can assess the features within the protein that indicate whether it is folded, unfolded or aggregated (Fig. 2). This comes from a close examination of not only the location and dispersion of the crosspeaks, but also the linewidths of the crosspeaks themselves. Changes in chemical shift, and in peak linewidths, allow structural and dynamical changes within a protein to be assessed. In addition, by incorporating the use of denaturants, pH or temperature, one can probe the unfolding refolding pathway of a protein at equilibrium. Indeed for proteins, such as lysozyme (16, 17) and alpha-lactalbumin (18, 19), the crosspeak distributions during unfolding revealed unique spectral qualities that were reminiscent of "molten globule" behaviour.

For proteins at high concentration, recording an HSQC is a rapid process (<30 min), particularly where a cryogenic probe is available. If sample is limited (e.g., ca. 10–20 μM), however, several hours may be required to obtain a spectrum with an

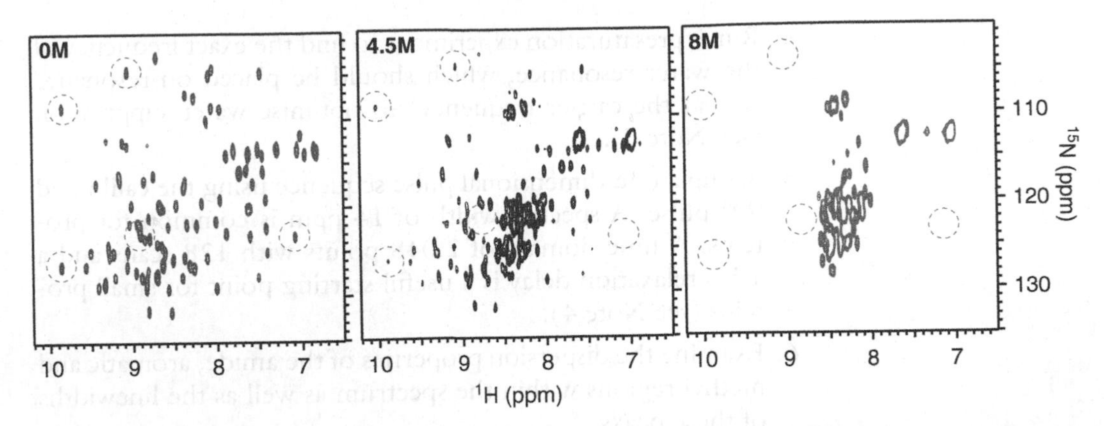

Fig. 2. The unfolding profile at equilibrium of a protein can be studied by observing the changes in chemical shift and the linewidth of resonances within ^{15}N-^{1}H HSQC spectra of samples incubated in increasing concentrations of urea. ^{15}N-^{1}H HSQC spectra of an Ig-like protein, ddFLN, incubated at 298 K in 0 M (*left*), 4.5 M (*middle*) and 8 M (*right*) urea is shown. During the unfolding of ddFLN, there is a progressive disappearance of dispersed resonances that are indicative of folded structure and the appearance of resonances that overlap between 8 and 8.5 ppm in the ^{1}H dimension, which is consistent with unfolded or disordered structure. The spectra also indicate that the resonances, a selection of which are highlighted by the circles, undergo fast exchange which results in the presence of a single resonance for each amino acid at any one time.

adequate signal to noise ratio using a spectrometer equipped with a cryoprobes. Recent advances in pulse sequence design have allowed for the development of rapid 2D spectra – in particular the SOFAST strategy which allows for decreased acquisition times. The SOFAST-HMQC relies on optimal and selective excitation of amide resonances, allowing the excited protons to relax (re-equilibrate) rapidly with the surrounding bath of unexcited "cold" protons (20). This relaxation or recovery time is a crucial factor in maximising the signal-to-noise ratio and is a trade-off between saving acquisition time and allowing the system to return closer to equilibrium. The SOFAST approach allows rapid repetition of data acquisition, and has been particularly useful in the examination of ribosome nascent chain complexes, which present a challenge to NMR where they are of low concentration and have a limited lifetime (21, 22).

NMR has historically been limited to the study of small (< ca. 25 kDa) proteins, due to the slow tumbling and hence rapid transverse relaxation and broad lines associated with larger molecules. However, this limit has been pushed back with the introduction and development of transverse relaxation optimised spectroscopy (TROSY) (23, 24), which has enabled NMR studies of large complexes, such as GroEL/ES (900 kDa) (8) and the proteasome (670 kDa) (9). Today, TROSY variants are available of most common biomolecular pulse programmes, and they require little additional calibration. The magnitude of the TROSY effect increases with the magnetic field strength (with a theoretical optimum for proton spectroscopy at a Larmor frequency of ca.

1,000 MHz (24); perdeuteration and selective labelling strategies are also generally required to eliminate undersirable relaxation pathways and simplify otherwise highly overlapped spectra (24).

The choice of temperature for protein NMR can be of great importance and depends strongly upon the folding state of the sample (discussed further in Subheading 3.6). Folded domains are most effectively observed at high temperatures, which maximises the tumbling of the molecules resulting in sharper resonances. While this is equally true for unfolded proteins, the additional effect of amide exchange must also be considered: at high temperatures, in the absence of secondary or tertiary structure, this can be quite rapid and result in the broadening of lines beyond detection (25). Therefore, unfolded or intrinsically disordered proteins are most effectively observed at lower temperatures.

3.5.1. Acquiring a ^{1}H-^{15}N HSQC Spectrum

1. Prepare a ^{15}N-labelled protein sample in 5–10% (v/v) D_2O at a minimum concentration of 100 μM (or greater if no cryoprobe is available) in a suitable buffer (see Note 1). Lower sample concentrations may be used, however, this leads to lower signal to noise and increases the acquisition time from minutes to several hours. This may not be conducive for unstable samples.
2. Equilibrate the sample at the desired temperature, then lock, tune and shim the spectrometer and find the 90° pulse.
3. Load a suitable HSQC pulse programme (see Note 6). To rapidly assess the quality of the ^{15}N signal arising from the sample and thus as, the first increment of a HSQC can be recorded, by setting a time domain of typically 1,024 complex points in the direct dimension, the collection of at least 16 scans (this depends on the sample concentration) but only a single point in the indirect dimension. This results in an ^{15}N edited 1D spectrum that can be used to optimise the experimental set up for a standard HSQC.
4. For a 2D, typical spectra are collected with 64 or 128 complex points in the indirect dimension (see Notes 4, 5 and 7). The number of scans required is dependent on the quality of the 1D and strongly upon the sample concentration. At a minimum, a complete phase cycle must be acquired, but for a good quality 2D spectrum it is not unusual to collect 16 to 32 scans. For a twofold increase in the signal-to-noise ratio of the protein, four times the number of scans is required (see Note 4). Initial spectral widths of 35 ppm (100–135 ppm) in the indirect (^{15}N) dimension and ca. 14 ppm in the direct (^{1}H) dimension should cover the dispersion profiles of a typical protein, and may later be optimised as required (see Note 7).
5. Set the receiver gain – this should be adjusted (manually or automatically) so that it is as high as possible without saturating the receiver.

6. The recycling (or relaxation) delay, typically 1–5 × T_1, can be adjusted to optimise the repetition rate of the experiment. A value of 1 s is usually adequate, but this should be increased if quantitative measurements of peak heights or volumes are required.

3.6. Interpretation of Protein HSQC Spectra

Protein HSQC spectra contain a great deal of structural and dynamical information on all the states a protein is populating. Folded, disordered and "intermediate" states all have certain recognisable characteristics which are discussed below. However, when multiple states are populated (e.g. at an unfolding midpoint), it is important to have an appreciation of the various ways in which chemical and conformational exchange can be manifested in NMR spectra. Key to this is the "NMR timescale" defined by the difference in frequency of the resonances, Δν (chemical shift difference multiplied by the Larmor frequency) of the order of magnitude of several milliseconds. If exchange between states occurs much more slowly than this ($k_{ex} << \Delta\nu$, "slow exchange"), the final spectrum will be a population-weighted superposition of the two components (exemplified by the urea unfolding of an immunoglobulin domain, Fig. 2). As the exchange rate increases and approaches Δν, crosspeaks broaden and shift towards each other, until at the coalesence point ($k_{ex} = 0.45\ \Delta\nu$, "intermediate exchange") the crosspeaks merge into a single, maximally broadened resonance – typically, broadened beyond detectability. As the exchange rate increases further, the line narrows, resulting in a single peak whose chemical shift is the population-weighted average of the original resonances. We therefore see that, depending upon the rate of folding and unfolding, protein denaturation as monitors by HSQC spectra may appear as a gradual movement of peaks from folded to unfolded chemical shifts (fast exchange); a uniform decrease of folded peaks concomitant with a uniform increase in unfolded peaks (slow exchange); or a broadening of all resonances at near the unfolding midpoint (intermediate exchange).

3.6.1. Native, Folded Protein

The HSQC spectrum of a folded protein generally exhibits well-resolved and broadly dispersed crosspeaks (Fig. 1b). Within the spectrum, certain amino acids groups have distinctive chemical shifts: for instance, glycine residues (^{1}H 8–8.5 ppm, ^{15}N 100–110 ppm), tryptophan sidechain indoles (^{1}H ca. 10.5 ppm, ^{15}N 125–135 ppm), asparagines/glutamine sidechain NH_2 groups (^{1}H 6.5–8 ppm, ^{15}N 110–115 ppm), arginine sidechain Hε (^{1}H 7–8.5 ppm, ^{15}N 87 ppm). Evaluating the number of crosspeaks and whether they correspond to the expected number of residues within the protein is an indication of whether there is heterogeneity within the sample, or in some instances, whether multiple conformations exist in slow exchange. When assessing the peaks, one must take into account that within a ^{1}H-^{15}N HSQC there is no proline crosspeak, while additional crosspeaks arise from the sidechains of tryptophan, asparagine, glutamine and arginine residues. Protein degradation also results in distinctive peaks in the "southeast" of the spectrum which may be a useful monitor of sample integrity.

An assessment of the linewidths is also an important consideration. For a well-folded protein, the expectation is that there is a uniform distribution of linewidths among the peaks. Regions of disorder may give rise to sharper peaks due to "motional narrowing" – unless the rate of amide hydrogen exchange with solvent water molecules is also increased, which may result in exchange broadening. Broadened crosspeaks may, however, also indicate residues undergoing chemical exchange, as a result of conformational fluctuations due to dynamics or weak protein–protein interactions, such as dimerisation. This may be tested by varying the sample concentration.

3.6.2. Unfolded/Intrinsically Disordered Proteins

The HSQC spectra of intrinsically disordered or unfolded proteins are easily distinguishable from folded states, with a narrow dispersion of crosspeaks within the ^{1}H dimension (8–8.5 ppm) (Fig. 2c). At low temperatures, crosspeaks are often sharper compared to those seen in folded states, but due to the influence of solvent–amide exchange this can be highly temperature dependent (25). Similar properties have been observed for other proteins, such as α-synuclein (26), Aβ peptides (27), cGMP phosphodiesterase (28), ZNF593 (29) and the C-terminal domain of HIV-1 Vif protein (30).

Where solvent exchange prevents the observation of amide resonances, and the sample conditions (e.g. temperature, pH) cannot be altered to compensate, the use of protonless NMR experiments can be explored. In such experiments, e.g. CON (correlating ^{13}C carbonyl shifts with the ^{15}N amide resonance) or similarly correlating carbonyl and alpha carbon shifts (COCA), ^{13}C magnetisation is directly detected, avoiding entirely the solvent exchange broadening associated with amide protons (31). However, due to the smaller gyromagnetic ratios of ^{13}C and ^{15}N compared to ^{1}H, the sensitivity of such experiments is significantly lower than proton-detected experiments, and long acquisition times (ca. 12–24 h) are required.

3.6.3. Protein-Folding Intermediates

During an equilibrium-folding/unfolding process, a protein may populate one or more intermediate states. The ability to probe a protein-folding intermediate depends on both its relative population compared to the folded and unfolded states and its structural properties. Many stable intermediates and "molten globule" states have been characterised biophysically and biochemically, and typically display features that show partial formation of secondary and/or tertiary structure, show hyperfluorescent properties upon interaction with the hydrophobic dye ANS, have expanded structures and are often susceptible to proteolysis (32, 33). Other "high-energy" intermediates are only ever populated transiently, and NMR relaxation methods can offer a sophisticated window into the structure and dynamics of these otherwise inaccessible states (34).

At the NMR level, protein-folding intermediates present a dynamic ensemble of states in rapid chemical exchange, whose conformations can fluctuate on the millisecond-microsecond timescale. As a result, the crosspeaks often suffer from severe line broadening, making interpretation of the spectra difficult. In some cases, the existence of an intermediate can only be demonstrated through the absence of folded and unfolded intensity (35, 36). In other cases, this broadening can be overcome to some degree, through a variation in denaturant/temperature/pH conditions. Particular success in identifying residue-specific information from molten globules has come from H/D exchange experiments (37) (Subheading 3.8). The alpha-lactalbumin molten globule, formed at low pH and exhibiting a highly broadened HSQC spectrum, has also been characterised with residue-specific resolution by gradual unfolding with denaturant and temperature, identifying the conditions under which a residue becomes observable (38, 39). Other examples of NMR-visible protein-folding intermediates come from studies of lysozyme (17), apomyglobin (40, 41) and apoflavodoxin (42), and in favourable cases measurements of NMR relaxation parameters (43) can be used to compare picosecond–nanosecond dynamics between intermediate and native states (18).

NMR can also be used to probe the structure and dynamics of high-energy intermediate states by measurements of relaxation dispersion, provided that the state is populated to at least ca. 0.5%, and exchange with the ground state occurs on the millisecond timescale (44, 45). The method, a detailed description of which is beyond the scope of this review, is based on the measurement of additional dephasing contributions (transverse relaxation) to ground state resonances arising from exchange with the transiently populated intermediate, and can provide information on the exchange rates, populations, and chemical shifts of the excited and otherwise "invisible" state. Such chemical shift differences have been used in conjunction with molecular dynamics simulations to describe in atomic detail the structure of an intermediate state ensemble in the Fyn SH3-folding pathway (34, 46), and also to demonstrate the coupling of folding of an intrinsically disordered transcription factor to the binding of its client (47). There is great potential for the use of excited state chemical shifts directly as restraints for structure determination algorithms (48, 49); additionally, by the measurement of spin-state selective relaxation dispersion in aligned media, residual dipolar couplings can be determined in the excited state, providing a particularly direct route to the structure determination of intermediate states (50).

3.7. Diffusion Measurements Reveal the Hydrodynamic Behaviour of a Protein

Proteins undergo two kinds of diffusion: rotational and translational. The *rotational* diffusion of a protein can be derived from relaxation experiments (51), which report on picosecond to nanosecond timescales of motion, and are discussed in the next section. *Translational* diffusion, on a millisecond to second timescale, can be measured using pulsed-field gradient (PFG) experiments (52, 53). These are particularly useful as they do not necessarily require isotopically labelled protein. Developed in the 1950s and 1960s, diffusion measurements have broad applicability, but in their simplest form can be used to correlate the hydrodynamic behaviour of a protein with its molecular weight and folding state (54). The physical basis of the method is that magnetic field gradients can be applied to spatially encode the position of a molecule (in a similar manner to MRI imaging). Some time (Δ) later, a reverse gradient is applied to decode the molecule's position; any movement due to diffusion results in a loss of signal intensity, I, according to the Stejskal-Tanner equation (52):

$$I = I_0 \exp[-\gamma^2 G^2 \delta^2 (\Delta - \delta/3) D] \qquad (1)$$

where I_0 is signal in the absence of gradients, $\gamma = 2.675 \times 10^8\ s^{-1} T^{-1}$ is the (proton) gyromagnetic ratio, G is the applied gradient strength, δ the gradient pulse length and Δ the diffusion period.

The use of diffusion measurements to examine protein folding and misfolding can provide information regarding the hydrodynamic behaviour during these events. It has also been particularly useful, for example, in studying the unfolded properties of the natively unfolded N-terminal domain of p53 where the diffusion coefficients were much smaller than indicated for a globular protein of a similar size (55, 56). PFG diffusion experiments have also been used recently to study local dynamics and complex formation in amyloid fibrils, which had been largely viewed as rigid, chemically inert structures (57, 58). This highlights the versatility of the strategy for examining protein misfolding in solution.

A variety of NMR experiments are available to measure diffusion, which differ in the precise manipulation of the spins involved. For most biomolecular purposes, we recommend the use of a stimulated echo sequence with bipolar gradients, which reduces distortions from eddy currents. The use of a sequence incorporating a longitudinal eddy current delay (LED) may be of benefit, although this is somewhat dependent upon the probe design.

3.7.1. Measurement of ^1H Diffusion

1. Prepare the sample and spectrometer as detailed in Protocols 3.1.1 and 3.4.1. It is recommended to include a low concentration (ca. 1 mM) of a small molecule, such as dioxane, which can act as an internal reference to compensate for any changes to the sample viscosity (54).
2. Set up a series of PFG experiments that acquire spectra with varying diffusion gradients, incrementing the z-gradient from

5 to 95% of the maximum value. The acquisition of at least 16 linearly spaced gradient increments is recommended.

3. The gradient pulse length (δ) and diffusion delay (Δ) must be optimised for a given protein, generally such that the signal is 95% attenuated at the maximum gradient strength. For typical biomolecular NMR probes, and small proteins, δ = 3 ms and Δ = 100 ms may provide an approximate starting point for further optimisation (see Note 8).
4. A wide spectral width of ca. 20 ppm is recommended to provide a large region for fitting and subtraction of the baseline.
5. Once the data has been acquired, dioxane (3.75 ppm) and protein methyl peaks should be integrated as functions of the gradient strength. Particular attention to phasing and baseline correction is required for accurate determination of diffusion coefficients.
6. If the gradient strength has been calibrated (for example, as described in (59)), integrated intensities may be fitted directly to Eq. 1 to determine the diffusion coefficient, *D*. Alternatively, dioxane and methyl intensities may be fitted as a function of the relative gradient strength, $G_{\%}$, to $I = I_0 \exp\left(-QG_{\%}^2\right)$. Given the Einstein–Stokes relation for the hydrodynamic radius, $r_h = kT/6\pi\eta D$ (where η is the viscosity), and that the hydrodynamic radius of dioxane is 2.12 Å (54), the hydrodynamic radius of the protein may be determined directly using Eq. 2:

$$r_h^{protein} = \frac{Q_{dioxane}}{Q_{protein}} r_h^{dioxane} \tag{2}$$

3.8. H/D Exchange to Probe Regions of Solvent Accessibility Within a Protein

A powerful means of examining protein dynamics is through the use of hydrogen/deuterium (H/D) exchange which, when combined with NMR, can reveal with residue-specific resolution the regions within a protein that are solvent exposed, both at equilibrium and along protein-folding and unfolding pathways. A simple theory of hydrogen exchange was described over half a century ago (60) and remains in popular use today (61). In this two-state model, a residue may be considered in either a closed state, in which the amide proton is protected and not susceptible to exchange; or, via an unspecified motion, an open state in which the amide proton is labile. From this open state, the residue may either return immediately to the closed and protected state, or it may exchange with a molecule of solvent water. The outcome of an opening event is therefore a competition between the closing rate, k_{close}, and the intrinsic, pseudo-random coil exchange rate, k_{int}.

The opening and closing rates are generally assumed to be independent of pH (in the absence of major structural changes); in contrast, as the exchange reaction is catalysed by both acid and base, the intrinsic exchange rate is strongly pH-dependent.

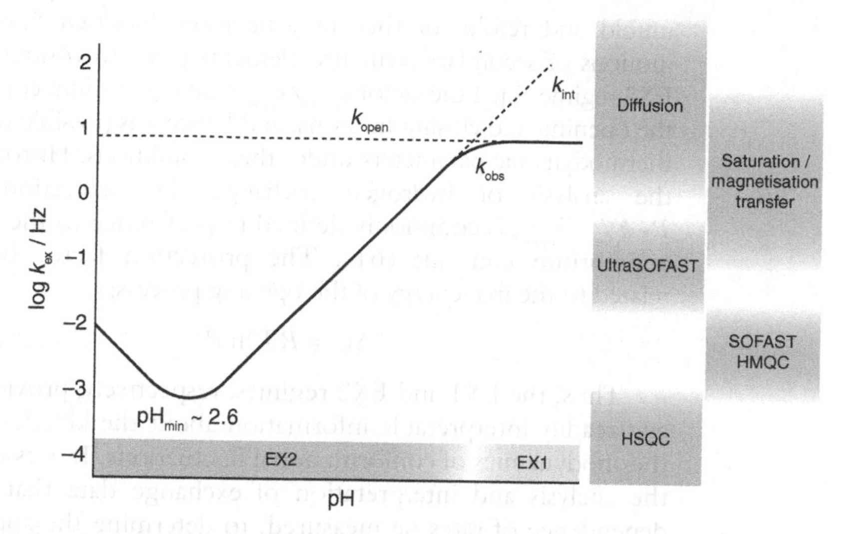

Fig. 3. The variation of the observed amide hydrogen exchange rate, k_{obs}, with solvent pH, showing the dependence upon the intrinsic "random coil" exchange rate, k_{int}, and the conformational opening rate, k_{open}, and the identification of the EX1 and EX2 exchange regimes. Also shown is a summary of applicable NMR methods for various timescales of exchange.

The observed exchange rate (and the nature of the rate-determining step) is therefore also dependent on the pH (Fig. 3) The identification of the rate-determining step, by determining the pH-dependence of the observed exchange rate, is of critical importance for the analysis and interpretation of hydrogen exchange data, and the labels EX1 and EX2 are used to describe the exchange regime at a particular pH. In the EX1 regime, intrinsic exchange is rapid ($k_{int} \gg k_{close}$), and the rate is limited by the opening rate. Hence, the observed rate k_{obs} is simply:

$$k_{obs}^{EX1} = k_{open} \tag{3}$$

Alternatively, if $k_{close} \gg k_{int}$ then there exists a pre-equilibrium between open and closed conformations prior to exchange. This is the EX2 limit of exchange, and the observed exchange rate may be expressed in terms of microscopic rate constants as:

$$k_{obs}^{EX2} = \frac{k_{open}}{k_{close}} k_{int} \tag{4}$$

The intrinsic exchange rate of an amide, k_{int}, may be calculated for a given pH to a generally acceptable degree of accuracy using tabulated values to compensate for the effect of neighbouring sidechains (a predominantly steric effect) (62); such calculations should also account for the temperature and isotope effects (63). Suitable algorithms for these calculations have been implemented in a convenient manner as the SPHERE Web service (64).

As demonstrated by Eq. 3, exchange in the EX1 regime yields information on the opening kinetics of structural fluctuations. Opening events may be global processes in which entire domains unfold and refold, or they may be more localised "breathing" motions of secondary structure elements (65). In contrast, in the EX2 regime Eq. 4 the factor k_{open}/k_{close} is an equilibrium constant for the opening and closing motions, and hence it is possible to extract thermodynamic parameters under these conditions. Historically, in the analysis of hydrogen exchange the protection factor, $P = k_{close}/k_{open}$, is commonly defined in preference to the forward equilibrium constant (61). The protection factor is readily related to the free energy of the opening process:

$$\Delta G = RT \ln P \quad (5)$$

Thus, the EX1 and EX2 regimes, respectively, provide useful and readily interpretable information about the kinetics and the thermodynamics of conformational fluctuations. It is essential for the analysis and interpretation of exchange data that the pH dependence of rates be measured, to determine the appropriate exchange regime for subsequent analyses (as per Fig. 3).

Protein H/D exchange is most commonly measured under "equilibrium" conditions, to investigate the high-energy unfolded or partially unfolded intermediates accessed from a given state. For residues involved in secondary structure elements within folded proteins, such exchange typically occurs on timescales of hours to months (66) and can be monitored in real time using ^{1}H-^{15}N HSQCs. Exchange on a minutes timescale may also be measured, using rapid acquisition strategies, such as the SOFAST experiments (20) discussed in Subheading 3, and exchange on timescales of seconds has been probed using the Ultra-SOFAST experiment in combination with stopped-flow apparatus (67). On shorter timescales still, a variety of relaxation-based methods may be employed to measure the kinetics of H/H exchange: this includes saturation transfer experiments (68), diffusion-based measurements (69, 70) and magnetisation transfer experiments, such as the CLEANEX-PM-FHSQC sequence (71). These are summarised in Fig. 3.

H/D exchange can also be employed in a pulsed manner, to probe the structure of intermediate states along a kinetic pathway. Depending on the unfolding and refolding timescales, this may require the use of stopped-flow mixing apparatus, (72) but this approach has also been used on slower timescales to probe the structure and molecular recycling of amyloid fibrils and protofibrils (73, 74).

3.8.1. Investigation of High-Energy Intermediates by Equilibrium H/D Exchange

1. Prepare two buffer solutions, one in 90% H_2O/10% D_2O, and the other in 100% D_2O, with identical and accurately calibrated pH(*) (see Note 9), containing 0.1% (w/v) DSS as an internal reference.
2. Lyophilize two aliquots of fully protonated protein, such that on resuspension in 600 μL of buffer, the concentration is suitable for rapid acquisition of HSQC spectra (on a cryogenic probe, at least 100 μM, but preferably 0.5–1 mM).

3. Resuspend one sample in 600 μL of the protonated buffer, and use to tune, shim and calibrate the spectrometer. Record a high-quality ^{1}H 1D and ^{1}H-^{15}N HSQC spectrum of the sample, which serves as the zero-time reference point. The sample should then be used to optimise the planned 2D experiments. Particular points to investigate are the number of complex points and the spectral width in the indirect (^{15}N) dimension, and the minimum number of scans required for phase cycling, which varies among different pulse programmes. Prepare a queue of these experiments sufficient to cover the required timescale; it is often helpful to gradually increase the number of scans collected at later times, reflecting the increasingly slow exchange and diminished signal intensity.
4. Remove the reference sample from the spectrometer, and empty the NMR tube: it is used for the exchanging sample to minimise the need for shimming. If exchange is being performed far from room temperature, pre-equilibrate the NMR tube and deuterated buffer at the required temperature.
5. Begin the exchange reaction by resuspending the second aliquot of protein in 600 μL of deuterated buffer, marking the exact time, and loading the sample into the spectrometer using the pre-shimmed NMR tube. Ensure that the sample is inserted to the same depth as before. Lock, and begin the queue of experiments immediately. Z1 and Z2 shims should be manually adjusted with reference to the lock signal, which also indicates when the sample has reached thermal equilibrium. At this point, abort the first experiment; the second queued experiment contains the first useable time point. With practice, and particularly with the help of a second person, dead times of ca. 2 min are achievable by this method.
6. On completion of the experimental session, record a ^{1}H 1D spectrum of the sample. This permits normalisation of DSS intensities, both with the initial reference point, and any later time-points that may be recorded. If such points are required (i.e. some amides remain unexchanged), the sample should be incubated at constant temperature throughout.
7. Peak heights or volumes should be fitted to a single exponential decay to determine the exchange rate. This procedure should be repeated at a minimum of two pH values to determine the exchange regime (EX1/EX2) for each residue.

3.8.2. Characterising On-Pathway Intermediate State Ensembles Using Pulsed H/D Exchange and NMR

1. The protein is unfolded by incubating in a concentrated deuterated denaturant (e.g. D_4-urea) in D_2O.
2. The protein is rapidly refolded into a D_2O-based folding buffer at relatively high pH (or low pH if the H exchange is slower than refolding).
3. After a chosen delay t_p, which depends on the time of folding, the protein is mixed with a high pH H_2O-based buffer. The time

t_p and pH of this "pulse" can be varied so that the exposed amides become fully protonated (NH) while structured and hence protected amides remain deuterated (see Note 10).

4. The D to H exchange reaction is quenched by lowering the pH, as the protein continues to fold to its native state. This process takes a snapshot of the H-D labelling that took place during the "pulse."
5. The sample is concentrated to a concentration suitable for NMR and a ^{1}H-^{15}N HSQC spectrum is recorded.
6. Control experiments which have samples at 0 and 100% labelling ensure that the crosspeak intensities can be calibrated.

3.9. Paramagnetic Relaxation Enhancement Experiments as a Probe of the Topology of a Protein

Paramagnetic relaxation experiments are reminiscent of FRET phenomenon that is used frequently in fluorescence spectroscopy. Its advantage over the NOESY is that it can probe inter-residue distances beyond 5 Å, and thus has been used to map long-range distances in the structure of T4 lysozyme (75), and first used to study protein folding in two independent studies, in the acid-denatured state of apomyglobin (76) in the urea-denatured state of staphylococcal nuclease (77, 78) and more recently, to study the long-range interactions present in the intrinsically disordered protein α-synuclein (79). In these cases, cysteine residues are exploited within the protein structure, to which a nitroxide spin-label, (e.g. MTSL) is attached. By subjecting the spin-label to its oxidised (paramagnetic) state, an enhancement in the relaxation properties in a ^{1}H-^{15}N HSQC is observed, as the unpaired electron is free to influence the magnetic properties of its neighbouring residues within a 10–25 Å radius; in the reduced (diamagnetic) state, however, this effect is not observed. The relaxation enhancement scales with the label's proximity as r^{-6}, and thus distance restraints can be derived. When combined with molecular dynamics simulations, this can generate an ensemble of structures that can be used to describe both the topology of a protein and the conformational space which it samples.

3.9.1. Preparation of Spin-Labelled Protein

1. Use site-directed mutagenesis techniques to introduce a single cysteine at a chosen location within the protein. Express and purify a suitably isotopically labelled version of the protein of interest.
2. Reduce the cysteine-variant protein by incubating with a tenfold molar excess of DTT for 1 h. Following this, remove the excess DTT by desalting (recommended: HiTrap Desalting column (GE Healthcare)).
3. Immediately following the DTT removal, add a tenfold molar excess of the spin label (e.g. MTSL, prepared as a 10 mg/ml stock in acetonitrile). Incubate the protein overnight (protected from light) at room temperature. Following the incubation, desalt the protein to remove the unreacted label.

3.9.2. Recording the Paramagnetic/Diamagnetic Properties of a Spin-Labelled Sample

1. Prepare a sample for NMR using step 1 in Protocol 3.1.
2. Record ^{1}H 1D and ^{1}H-^{15}N HSQC (Protocols 3.4 and 3.5). This ensures that the presence of the label has not affected the overall integrity of the protein.
3. Record an HSQC (Protocol 3.5.1) of the sample, this represents the oxidised (paramagnetic) state.
4. To the oxidised state, add a 3 M excess of sodium ascorbate (use a concentrated stock to ensure that there are minimal effects on the sample's pH/volume).
5. Record an HSQC (Protocol 3.5.1) of the sample using 16 scans per increment, with 1024 complex points in the direct dimension and 128 complex points in the indirect dimension, this represents the reduced (diamagnetic) state.
6. Record a control samples of labelled protein and also unlabelled (with MTSL attached) in the presence of a fivefold molar excess of sodium ascorbate. This detects any intermolecular interactions at high sensitivity that would appear as reductions in the crosspeak intensity.

3.10. Summary

Studying protein folding and misfolding by NMR is a powerful means of providing a residue-specific level of detail of structure, dynamics and conformational change. NMR spectroscopy continues to evolve and many leaps have been made in studying large systems, such as the 26S proteosome and the ribosome, that have been traditionally considered unsuitable for NMR. Moreover, the ability to study real-time folding and in-cell folding events are becoming more commonplace with the field and it opens up new avenues for study. Importantly, however, all of these approaches are taking protein folding a step closer to examining protein folding/misfolding as it would occur within the cell.

4. Notes

1. Buffers: Use a buffer that keeps the protein soluble while having a low ionic strength. High ionic strength decreases the coupling of the sample to the probe, resulting in longer pulse lengths, higher power levels, reduced sensitivity and therefore longer acquisition times. Phosphate is often a good buffer for NMR studies because it does not contain any protons which might overlap with protein resonances (although it has a relatively high ionic strength). Other buffers such as HEPES and Tris are acceptable, however, and are available in deuterated forms (at significantly greater expense) if required.
2. DSS: This is an IUPAC-recognised standard for the calibration of chemical shifts in water, having a strong resonance at 0 ppm.

Three smaller peaks are observed at 0.8, 1.9 and 3.1 ppm, although these can be suppressed if a deuterated form of the compound is used. The DSS resonance is also useful as an intensity reference when performing titrations and similar measurements. For convenience, DSS can also be purchased as a 1% (w/v) solution in D_2O.

3. Water suppression: Due to the high proton concentration of water (110 M), good water suppression is critical to obtaining high-quality spectra. Residual water can obscure protein resonances and limits the gain (sensitivity) that can be implemented, which leads to an overall poorer detection of the protein's signal. The quality of the shim strongly influences water suppression, as can the choice of pulse sequence. For a 1D spectrum, pulse programmes using WATERGATE or excitation sculpting sequences are recommended (80).
4. Determining the number of scans and complex points to collect for a given protein is an empirical process and depends heavily on both the concentration and also on the relaxation properties of the protein. A higher concentration and more scans both lead to an improved signal/noise ratio, although it is more efficient to increase the concentration, as signal/noise scales linearly with the concentration, but only as the square root of the number of scans; i.e. doubling the protein concentration doubles the signal/noise ratio, but to achieve a similar increase with the same sample would require four times the number of scans. If the signal (Free Induction Decay, FID) from a protein decays (relaxes) too quickly, then trying to accumulate too many points within a given scan may result only in the acquisition of noise.
5. There is a trade off between the number of points collected in the indirect (^{15}N or ^{13}C) dimensions and the number of scans that is recorded. These are all optimised in combination to maximise the signal to noise ratio in the sample. For each successive complex point, there is an incremental increase in the evolution time within the pulse sequence. An empirical understanding of the relaxation properties of the protein is required to ensure that additional points improve the overall signal arising from the protein rather than acquiring noise due to the loss of signal through relaxation pathways.
6. A large variety of HSQC pulse sequences are available. Efficient water suppression is of great importance; a combination of a WATERGATE filter with a water flip-back pulse is effective for most purposes (80, 81). "Sensitivity-enhanced" experiments (82, 83) theoretically increase the signal-to-noise ratio by a factor of $\sqrt{2}$, albeit using a longer pulse sequence which increases losses to relaxation; the utility of this method should be tested on a protein-by-protein basis. When working with large proteins (> ca. 25 kDa), TROSY variants are essential.

7. To maximise data collection, a careful consideration of the spectral width in the indirect dimension is recommended. Record a "quick" 2D, and examine the dispersion of peaks. The difference between the two most dispersed peaks represents the spectral width.
8. For experiments using bipolar gradients, the gradient pulse length is often specified as $\delta/2$. Care should be taken not to overload the gradient amplifier.
9. pH* is the uncorrected reading of deuterated solution with an H_2O-calibrated pH electrode, and may be converted to pD by the addition of 0.4 (84).
10. By setting up a series of titrations, the protein can be unfolded/refolded into different final concentrations of urea and the un/folding pathway of a protein can be examined.

Acknowledgements

We would like to thank the members of the Christodoulou lab for useful discussions and Dr. John Kirkpatrick for critical reading of this chapter.

References

1. Yamazaki, T., R. Muhandiram, and L.E. Kay. (1994) NMR experiments for the measurement of carbon relaxation properties in highly enriched, uniformly 13C, 15N-labeled proteins: application to 13C alpha carbons. *J Am Chem Soc.* **116**, 8266–8278.
2. Markus, M.A., K.T. Dayie, P. Matsudaira, and G. Wagner. (1994) Effect of deuteration on the amide proton relaxation rates in proteins. Heteronuclear NMR experiments on villin 14T. *J Magn Reson B.* **105**, 192–195.
3. Kobayashi, M., H. Yagi, T. Yamazaki, M. Yoshida, and H. Akutsu. (2008) Dynamic inter-subunit interactions in thermophilic F(1)-ATPase subcomplexes studied by cross-correlated relaxation-enhanced polarization transfer NMR. *J Biomol NMR.* **40**, 165–74.
4. Mittermaier, A. and L.E. Kay. (2002) Effect of deuteration on some structural parameters of methyl groups in proteins as evaluated by residual dipolar couplings. *J Biomol NMR.* **23**, 35–45.
5. Vasos, P.R., J.B. Hall, R. Kummerle, and D. Fushman. (2006) Measurement of 15N relaxation in deuterated amide groups in proteins using direct nitrogen detection. *J Biomol NMR.* **36**, 27–36.
6. LeMaster, D.M. (1990) Uniform and selective deuteration in two-dimensional NMR of proteins. *Annu Rev Biophys Biophys Chem.* **19**, 243–66.
7. Hwang, K.J., F. Mahmoodian, J.A. Ferretti, E.D. Korn, and J.M. Gruschus. (2007) Intramolecular interaction in the tail of Acanthamoeba myosin IC between the SH3 domain and a putative pleckstrin homology domain. *Proc Natl Acad Sci U S A.* **104**, 784–9.
8. Fiaux, J., E.B. Bertelsen, A.L. Horwich, and K. Wüthrich. (2002) NMR analysis of a 900K GroEL GroES complex. *Nature.* **418**, 207–11.
9. Sprangers, R. and L.E. Kay. (2007) Quantitative dynamics and binding studies of the 20S proteasome by NMR. *Nature.* **445**, 618–22.
10. Sprangers, R., A. Gribun, P.M. Hwang, W.A. Houry, and L.E. Kay. (2005) Quantitative NMR spectroscopy of supramolecular complexes: dynamic side pores in ClpP are important for product release. *Proc Natl Acad Sci U S A.* **102**, 16678–83.
11. Artero, J.B., M. Hartlein, S. McSweeney, and P. Timmins. (2005) A comparison of refined X-ray structures of hydrogenated and perdeuterated rat gammaE-crystallin in H2O and D2O. *Acta Crystallogr D Biol Crystallogr.* **61**, 1541–9.

12. Tugarinov, V., V. Kanelis, and L.E. Kay. (2006) Isotope labeling strategies for the study of high-molecular-weight proteins by solution NMR spectroscopy. *Nat Protoc.* **1**, 749–54.
13. McIntosh, L.P. and F.W. Dahlquist. (1990) Biosynthetic incorporation of 15N and 13C for assignment and interpretation of nuclear magnetic resonance spectra of proteins. *Q Rev Biophys.* **23**, 1–38.
14. Whittaker, J.W. (2007) Selective isotopic labeling of recombinant proteins using amino acid auxotroph strains. *Methods Mol Biol.* **389**, 175–88.
15. Cavanagh, J., *Protein NMR spectroscopy: principles and practice.* 2nd ed. 2007: Academic Press.
16. Radford, S.E., C.M. Dobson, and P.A. Evans. (1992) The folding of hen lysozyme involves partially structured intermediates and multiple pathways. *Nature.* **358**, 302–7.
17. Miranker, A., S.E. Radford, M. Karplus, and C.M. Dobson. (1991) Demonstration by NMR of folding domains in lysozyme. *Nature.* **349**, 633–6.
18. Redfield, C., R.A. Smith, and C.M. Dobson. (1994) Structural characterization of a highly-ordered "molten globule" at low pH. *Nat Struct Biol.* **1**, 23–9.
19. Wijesinha-Bettoni, R., C.M. Dobson, and C. Redfield. (2001) Comparison of the denaturant-induced unfolding of the bovine and human alpha-lactalbumin molten globules. *J Mol Biol.* **312**, 261–73.
20. Schanda, P., E. Kupce, and B. Brutscher. (2005) SOFAST-HMQC experiments for recording two-dimensional heteronuclear correlation spectra of proteins within a few seconds. *J Biomol NMR.* **33**, 199–211.
21. Hsu, S.T., P. Fucini, L.D. Cabrita, H. Launay, C.M. Dobson, and J. Christodoulou. (2007) Structure and dynamics of a ribosome-bound nascent chain by NMR spectroscopy. *Proc Natl Acad Sci U S A.* **104**, 16516–21.
22. Cabrita, L.D., S.T. Hsu, H. Launay, C.M. Dobson, and J. Christodoulou. (2009) Probing ribosome-nascent chain complexes produced in vivo by NMR spectroscopy. *Proc Natl Acad Sci U S A.* **106**, 22239–22244.
23. Pervushin, K., R. Riek, G. Wider, and K. Wuthrich. (1997) Attenuated T2 relaxation by mutual cancellation of dipole-dipole coupling and chemical shift anisotropy indicates an avenue to NMR structures of very large biological macromolecules in solution. *Proc Natl Acad Sci U S A.* **94**, 12366–71.
24. Fernandez, C. and G. Wider. (2003) TROSY in NMR studies of the structure and function of large biological macromolecules. *Curr Opin Struct Biol.* **13**, 570–80.
25. Croke, R.L., C.O. Sallum, E. Watson, E.D. Watt, and A.T. Alexandrescu. (2008) Hydrogen exchange of monomeric alpha-synuclein shows unfolded structure persists at physiological temperature and is independent of molecular crowding in Escherichia coli. *Protein Sci.* **17**, 1434–45.
26. Eliezer, D., E. Kutluay, R. Bussell, Jr., and G. Browne. (2001) Conformational properties of alpha-synuclein in its free and lipid-associated states. *J Mol Biol.* **307**, 1061–73.
27. Macao, B., W. Hoyer, A. Sandberg, A.C. Brorsson, C.M. Dobson, and T. Hard. (2008) Recombinant amyloid beta-peptide production by coexpression with an affibody ligand. *BMC Biotechnol.* **8**, 82.
28. Song, J., L.W. Guo, H. Muradov, N.O. Artemyev, A.E. Ruoho, and J.L. Markley. (2008) Intrinsically disordered gamma-subunit of cGMP phosphodiesterase encodes functionally relevant transient secondary and tertiary structure. *Proc Natl Acad Sci U S A.* **105**, 1505–10.
29. Hayes, P.L., B.L. Lytle, B.F. Volkman, and F.C. Peterson. (2008) The solution structure of ZNF593 from Homo sapiens reveals a zinc finger in a predominantly unstructured protein. *Protein Sci.* **17**, 571–6.
30. Reingewertz, T.H., H. Benyamini, M. Lebendiker, D.E. Shalev, and A. Friedler. (2009) The C-terminal domain of the HIV-1 Vif protein is natively unfolded in its unbound state. *Protein Eng Des Sel.* **22**, 281–7.
31. Bermel, W., I. Bertini, I.C. Felli, M. Piccioli, and R. Pierattelli. (2005) 13C-detected protonless NMR spectroscopy of proteins in solution. *Prog Nucl Magn Res Sp.* **48**, 25–45.
32. Arai, M. and K. Kuwajima. (2000) Role of the molten globule state in protein folding. *Adv Protein Chem.* **53**, 209–82.
33. Ptitsyn, O.B. (1995) Molten globule and protein folding. *Adv Protein Chem.* **47**, 83–229.
34. Korzhnev, D.M., X. Salvatella, M. Vendruscolo, A.A. Di Nardo, A.R. Davidson, C.M. Dobson, and L.E. Kay. (2004) Low-populated folding intermediates of Fyn SH3 characterized by relaxation dispersion NMR. *Nature.* **430**, 586–590.
35. Hsu, S.-T.D., L.D. Cabrita, P. Fucini, C.M. Dobson, and J. Christodoulou. (2009) Structure, dynamics and folding of an immunoglobulin domain of the gelation factor (ABP-120) from Dictyostelium discoideum. *J Mol Biol.* **388**, 865–79.
36. Garcia, P., L. Serrano, M. Rico, and M. Bruix. (2002) An NMR view of the folding process of a CheY mutant at the residue level. *Structure.* **10**, 1173–1185.
37. Redfield, C. (2004) Using nuclear magnetic resonance spectroscopy to study molten globule states of proteins. *Methods.* **34**, 121–32.

38. Quezada, C.M., B.A. Schulman, J.J. Froggatt, C.M. Dobson, and C. Redfield. (2004) Local and global cooperativity in the human alpha-lactalbumin molten globule. *J Mol Biol.* **338**, 149–58.
39. Schulman, B.A., P.S. Kim, C.M. Dobson, and C. Redfield. (1997) A residue-specific NMR view of the non-cooperative unfolding of a molten globule. *Nat Struct Biol.* **4**, 630–4.
40. Uzawa, T., C. Nishimura, S. Akiyama, K. Ishimori, S. Takahashi, H.J. Dyson, and P.E. Wright. (2008) Hierarchical folding mechanism of apomyoglobin revealed by ultra-fast H/D exchange coupled with 2D NMR. *Proc Natl Acad Sci U S A.* **105**, 13859–64.
41. Hughson, F.M., P.E. Wright, and R.L. Baldwin. (1990) Structural characterization of a partly folded apomyoglobin intermediate. *Science.* **249**, 1544–8.
42. van Mierlo, C.P., J.M. van den Oever, and E. Steensma. (2000) Apoflavodoxin (un)folding followed at the residue level by NMR. *Protein Sci.* **9**, 145–57.
43. Jarymowycz, V.A. and M.J. Stone. (2006) Fast time scale dynamics of protein backbones: NMR relaxation methods, applications, and functional consequences. *Chem Rev.* **106**, 1624–71.
44. Neudecker, P., P. Lundstrom, and L.E. Kay. (2009) Relaxation dispersion NMR spectroscopy as a tool for detailed studies of protein folding. *Biophys J.* **96**, 2045–54.
45. Hansen, D.F., P. Vallurupalli, P. Lundstrom, P. Neudecker, and L.E. Kay. (2008) Probing chemical shifts of invisible states of proteins with relaxation dispersion NMR spectroscopy: how well can we do? *J Am Chem Soc.* **130**, 2667–75.
46. Neudecker, P., A. Zarrine-Afsar, A.R. Davidson, and L.E. Kay. (2007) Phi-value analysis of a three-state protein folding pathway by NMR relaxation dispersion spectroscopy. *Proc Natl Acad Sci U S A.* **104**, 15717–22.
47. Sugase, K., H.J. Dyson, and P.E. Wright. (2007) Mechanism of coupled folding and binding of an intrinsically disordered protein. *Nature.*
48. Cavalli, A., X. Salvatella, C.M. Dobson, and M. Vendruscolo. (2007) Protein structure determination from NMR chemical shifts. *Proc Natl Acad Sci U S A.* **104**, 9615–20.
49. Shen, Y., O. Lange, F. Delaglio, P. Rossi, J.M. Aramini, G. Liu, A. Eletsky, Y. Wu, K.K. Singarapu, A. Lemak, A. Ignatchenko, C.H. Arrowsmith, T. Szyperski, G.T. Montelione, D. Baker, and A. Bax. (2008) Consistent blind protein structure generation from NMR chemical shift data. *Proc Natl Acad Sci U S A.* **105**, 4685–90.
50. Vallurupalli, P., D.F. Hansen, E. Stollar, E. Meirovitch, and L.E. Kay. (2007) Measurement of bond vector orientations in invisible excited states of proteins. *Proc Natl Acad Sci U S A.* **104**, 18473–7.
51. Price, W.S., A.V. Barzykin, K. Hayamizu, and M. Tachiya. (1998) A model for diffusive transport through a spherical interface probed by pulsed-field gradient NMR. *Biophys J.* **74**, 2259–71.
52. Stejskal, E.O. and J.E. Tanner. (1965) Spin diffusion measurements: Spin echoes in the presence of a time-dependent field gradient. *J. Chem. Phys.* **42**, 288–292.
53. Palmer, A.G., 3rd. (1997) Probing molecular motion by NMR. *Curr Opin Struct Biol.* **7**, 732–7.
54. Wilkins, D.K., S.B. Grimshaw, V. Receveur, C.M. Dobson, J.A. Jones, and L.J. Smith. (1999) Hydrodynamic radii of native and denatured proteins measured by pulse field gradient NMR techniques. *Biochemistry.* **38**, 16424–16431.
55. Dehner, A. and H. Kessler. (2005) Diffusion NMR spectroscopy: folding and aggregation of domains in p53. *Chembiochem.* **6**, 1550–65.
56. Dawson, R., L. Muller, A. Dehner, C. Klein, H. Kessler, and J. Buchner. (2003) The N-terminal domain of p53 is natively unfolded. *J Mol Biol.* **332**, 1131–41.
57. Baldwin, A.J., J. Christodoulou, P.D. Barker, C.M. Dobson, and G. Lippens. (2007) Contribution of rotational diffusion to pulsed field gradient diffusion measurements. *J Chem Phys.* **127**, 114505.
58. Waudby, C.A., T.P. Knowles, G.L. Devlin, J.N. Skepper, H. Ecroyd, J.A. Carver, M.E. Welland, J. Christodoulou, C.M. Dobson, and S. Meehan. (2010) The interaction of alphaB-crystallin with mature alpha-synuclein amyloid fibrils inhibits their elongation. *Biophys J.* **98**, 843–51.
59. Berger, S. and S. Braun, *200 and more NMR experiments: A practical course.* 3rd ed. 2004: Wiley-VCH.
60. Linderstrom-Lang, K., ed. *Deuterium exchange and protein structure.* Symposium on protein structure, ed. A. Neuberger. 1958, Methuen: London
61. Dempsey, C. (2001) Hydrogen exchange in peptides and proteins using NMR spectroscopy. *Prog Nucl Magn Res Sp* **39**, 135–170.
62. Bai, Y., J.S. Milne, L. Mayne, and S.W. Englander. (1993) Primary structure effects on peptide group hydrogen exchange. *Proteins.* **17**, 75–86.
63. Connelly, G.P., Y. Bai, M.F. Jeng, and S.W. Englander. (1993) Isotope effects in peptide group hydrogen exchange. *Proteins.* **17**, 87–92.

64. Zhang, Y. A server program for hydrogen exchange rate estimation (http://www.fccc.edu/research/labs/roder/sphere/sphere.html).
65. Englander, S.W., T.R. Sosnick, J.J. Englander, and L. Mayne. (1996) Mechanisms and uses of hydrogen exchange. *Curr Opin Struct Biol.* **6**, 18–23.
66. Huang, J.R., T.D. Craggs, J. Christodoulou, and S.E. Jackson. (2007) Stable intermediate states and high energy barriers in the unfolding of GFP. *J Mol Biol.* **370**, 356–71.
67. Gal, M., P. Schanda, B. Brutscher, and L. Frydman. (2007) UltraSOFAST HMQC NMR and the repetitive acquisition of 2D protein spectra at Hz rates. *J Am Chem Soc.* **129**, 1372–7.
68. Spera, S., M. Ikura, and A. Bax. (1991) Measurement of the exchange rates of rapidly exchanging amide protons: application to the study of calmodulin and its complex with a myosin light chain kinase fragment. *J Biomol NMR.* **1**, 155–65.
69. Andrec, M. and J.H. Prestegard. (1997) Quantitation of chemical exchange rates using pulsed-field-gradient diffusion measurements. *J Biomol NMR.* **9**, 136–50.
70. Bockmann, A. and E. Guittet. (1997) Determination of fast proton exchange rates of biomolecules by NMR using water selective diffusion experiments. *FEBS Lett.* **418**, 127–30.
71. Hwang, T.L., P.C. van Zijl, and S. Mori. (1998) Accurate quantitation of water-amide proton exchange rates using the phase-modulated CLEAN chemical EXchange (CLEANEX-PM) approach with a Fast-HSQC (FHSQC) detection scheme. *J Biomol NMR.* **11**, 221–6.
72. Bollen, Y.J., M.B. Kamphuis, and C.P. van Mierlo. (2006) The folding energy landscape of apoflavodoxin is rugged: hydrogen exchange reveals nonproductive misfolded intermediates. *Proc Natl Acad Sci U S A.* **103**, 4095–100.
73. Carulla, N., G.L. Caddy, D.R. Hall, J. Zurdo, M. Gairi, M. Feliz, E. Giralt, C.V. Robinson, and C.M. Dobson. (2005) Molecular recycling within amyloid fibrils. *Nature.* **436**, 554–8.
74. Carulla, N., M. Zhou, M. Arimon, M. Gairi, E. Giralt, C.V. Robinson, and C.M. Dobson. (2009) Experimental characterization of disordered and ordered aggregates populated during the process of amyloid fibril formation. *Proc Natl Acad Sci U S A.* **106**, 7828–33.
75. Voss, J., L. Salwinski, H.R. Kaback, and W.L. Hubbell. (1995) A method for distance determination in proteins using a designed metal ion binding site and site-directed spin labeling: evaluation with T4 lysozyme. *Proc Natl Acad Sci U S A.* **92**, 12295–9.
76. Lietzow, M.A., M. Jamin, H.J. Jane Dyson, and P.E. Wright. (2002) Mapping long-range contacts in a highly unfolded protein. *J Mol Biol.* **322**, 655–62.
77. Gillespie, J.R. and D. Shortle. (1997) Characterization of long-range structure in the denatured state of staphylococcal nuclease. I. Paramagnetic relaxation enhancement by nitroxide spin labels. *J Mol Biol.* **268**, 158–69.
78. Gillespie, J.R. and D. Shortle. (1997) Characterization of long-range structure in the denatured state of staphylococcal nuclease. II. Distance restraints from paramagnetic relaxation and calculation of an ensemble of structures. *J Mol Biol.* **268**, 170–84.
79. Dedmon, M.M., K. Lindorff-Larsen, J. Christodoulou, M. Vendruscolo, and C.M. Dobson. (2005) Mapping long-range interactions in alpha-synuclein using spin-label NMR and ensemble molecular dynamics simulations. *J Am Chem Soc.* **127**, 476–7.
80. Piotto, M., V. Saudek, and V. Sklenar. (1992) Gradient-tailored excitation for single-quantum NMR spectroscopy of aqueous solutions. *J Biomol NMR.* **2**, 661–5.
81. Grzesiek, S. and A. Bax. (1993) Measurement of amide proton exchange rates and NOEs with water in 13C/15N-enriched calcineurin B. *J Biomol NMR.* **3**, 627–38.
82. Palmer, A.G., 3rd, J. Cavanagh, P.E. Wright, and M. Rance. (1991) Sensitivity improvement in proton-detected two-dimensional heteronuclear correlation NMR spectroscopy. *J Magn Reson.* **93**, 151–170.
83. Cavanagh, J., A.G. Palmer, 3rd, P.E. Wright, and M. Rance. (1991) Sensitivity improvement in proton-detected two-dimensional heteronuclear relay spectoscopy. *J Magn Reson.* **91**, 429–436.
84. Krezel, A. and W. Bal. (2004) A formula for correlating pKa values determined in D2O and H2O. *J Inorg Biochem.* **98**, 161–6.

Chapter 8

Diagnostics for Amyloid Fibril Formation: Where to Begin?

Danny M. Hatters and Michael D.W. Griffin

Abstract

Twenty-five proteins are known to form amyloid fibrils in vivo in association with disease (Westermark et al., Amyloid 12:1–4, 2005). However, the fundamental ability of a protein to form amyloid-like fibrils is far more widespread than in just the proteins associated with disease, and indeed this property can provide insight into the basic thermodynamics of folding and misfolding pathways. But how does one determine whether a protein has formed amyloid-like fibrils? In this chapter, we cover the basic steps toward defining the amyloid-like properties of a protein and how to measure the kinetics of fibrillization. We describe several basic tests for aggregation and the binding to two classic amyloid-reactive dyes, Congo Red, and thioflavin T, which are key indicators to the presence of fibrils.

Key words: Protein misfolding, Congo red, Thioflavin T, Aggregation, Birefringence, Amyloid, Fibril, Procedures, Protocol, Method

1. Introduction

When considering whether a protein is an "amyloid" or not, it is worth first considering the historical context of "amyloid." Amyloid, derived from the latin word *amylum* for "starch," was first described as iodine-reactive deposits in the brain (1). While the iodine reactivity likely reflects proteoglycan staining rather than starch, which is not present in human tissue, later studies described amyloid deposits as being red–green birefringent under polarized light in the presence of the stain Congo Red (2, 3). Ex vivo amyloid deposits were shown to contain fibrous material, which confers the Congo Red birefringent properties observed under light microscopy (4). From these and other studies, the definition for "amyloid" evolved as a histological description for extracellular deposits exhibiting Congo Red red–green birefringence comprising primarily of one or more of 25 particular proteins (5).

Andrew F. Hill et al. (eds.), *Protein Folding, Misfolding, and Disease: Methods and Protocols*, Methods in Molecular Biology, vol. 752, DOI 10.1007/978-1-60327-223-0_8, © Springer Science+Business Media, LLC 2011

Since the mid 1990s, much insight has been gleaned on the molecular structure and mechanisms of the formation of amyloid fibrils (6). While the histological definition of "amyloid" does not directly translate to the biochemical and biophysical properties of the material, a number of structural attributes common to the proteins in amyloid deposits have been identified. These include an aggregate assembly state of fibrillar morphology rich in β-sheet secondary structure comprising β-strands aligned perpendicular to the long fibril axis (7–9). The amyloid structure for many proteins, such as lysozyme, was shown to reflect a nonnative alternative conformation to the native state, suggesting amyloid depicts the consequences of protein misfolding in vivo (10). Hence, exploring how environmental parameters or mutations mediate fibril formation can be useful for understanding the folding thermodynamics of a protein.

The biophysical and structural properties of amyloid proteins have more recently been shown to occur in a large number of proteins other than those classically defined by pathology, which has led to the suggestion that proteins are inherently capable of acquiring an amyloid-like structure under certain conditions (11). Indeed, some bacterial and fungal proteins seem to have evolved specifically as natural "amyloid" structure for normal function (12, 13).

In this chapter, we describe the first steps in defining an amyloid fibril from a biochemical perspective. We anticipate that the approaches described here be the starting point for experimentalists and feed directly to the other chapters of this book for a more detailed analysis. We cover three basic spectroscopic approaches: direct assessment of aggregation, and the binding to two classic amyloid dyes Congo Red and thioflavin T (ThT), and finish with procedures for defining the kinetics of amyloid–fibril formation.

2. Materials

2.1. Measuring Aggregation State by Centrifugation

1. Microcentrifuge.
2. A high-speed benchtop centrifuge (see Note 1).
3. Sample buffer, e.g., phosphate-buffered saline (PBS) (see Note 2).
4. UV–Vis Spectrophotometer.
5. Protein detection reagents (see Note 3).

2.2. Measuring Aggregation State by Gel Filtration Chromatography

1. HPLC system/peristaltic pump (if available) and chromatography column (e.g., Superdex 200 or Sepharose CL-4B from GE Healthcare).

2. Sample buffer, e.g., PBS. When using HPLC systems, all buffers must be filtered through a 0.22 μm filter to remove particulates. Always prepare fresh buffers to prevent mold contamination (Refer also to Note 2).

2.3. Measuring Congo Red Reactivity and Birefringence

1. Congo Red for the spectroscopic assay. Prepare a 1 mM solution in ethanol (Congo Red is poorly soluble in water).
2. Congo Red for the birefringence assay. Add a saturating amount of Congo Red to a small volume of 80% (v/v) ethanol, 20% (v/v) distilled water, and saturated NaCl. Stir the solution for 10 min and allow the undissolved Congo Red to settle. Filter the supernatant through a 0.22 μm syringe filter to obtain the working solution. This solution can be stored for a number of days if kept in the dark but should be filtered directly before use.
3. Microfiltration centrifugation device. Any of the standard brands is fine. It is best to use a high nominal molecular weight cut-off (NMWCO), which facilitates rapid concentration without the loss of fibrillar material relative to the low NMWCO.
4. Light microscope equipped with incident light and objective polarizers (for example, those commonly used to examine protein crystals).
5. A spectrophotometer capable of wavelength scans in the wavelength range 300–700 nm.
6. Sample buffer, e.g., PBS (Refer also to Note 2).

2.4. Measuring Thioflavin T Reactivity

1. Prepare a 1 mM stock ThT solution by dissolving the appropriate amount of dye in PBS, and filter the solution through a 0.22 μm syringe filter. ThT solutions must be stored in the dark and are stable at room temperature under these conditions for at least one week. Solutions can also be stored long term at −20°C in the dark for a number of months.
2. Fluorimeter. Any standard instrument is fine. A plate reader format is most useful for kinetic studies.
3. Sample buffer, e.g., PBS (Refer also to Note 2).

2.5. Kinetic Assays

1. Refer to materials and buffers needed Subheadings 2.1–2.4.
2. Mineral oil.

3. Methods

3.1. Measuring Aggregation State by Centrifugation

Amyloid fibrils are characterized by their large molecular mass relative to the “normal” form of the protein. Hence, centrifugation remains a simple and direct test for the solubility of the protein.

Often large fibrils appear flocculent in solution when viewed through a microfuge tube against bright light. Such fibrils are easily pelleted in a microcentrifuge and can be readily measured for solubility as follows:

1. Prepare two parallel samples of the protein in two 1.5 ml microcentrifuge tubes. At least 200 μl of sample is needed in one of the tubes for centrifugation, and 60 μl of sample is needed in the second tube for the noncentrifuged sample.
2. For the 200 μl sample, centrifuge for 30 min at room temperature at maximum speed in a microcentrifuge. See Note 4.
3. Carefully remove and keep aside 50 μl of the supernatant without disturbing the pellet.
4. Measure the protein concentration of the supernatant versus the uncentrifuged sample. It is important to thoroughly mix the uncentrifuged sample before analysis to ensure that the aggregates have not settled. The protein detection can be done a number of ways and adapted to a 96-well plate for use in a plate reader (see Note 3).
   It is important to note that not all fibrils (for example, those formed by apoC-II (14) or apoE (15)) are visibly flocculent or pellet in a microcentrifuge. In addition, small oligomer precursors to large fibrils may also remain in solution. Hence, to comprehensively capture the oligomeric, aggregated forms of the protein, pelleting should also be investigated using a high-speed benchtop centrifuge (such as a Beckman Optima TLX ultracentrifuge).
5. Prepare aliquots of samples as described above in step 1.
6. Centrifuge the 200 μl aliquot at 100,000 × *g* for 30 min using a high-speed fixed angle benchtop centrifuge (for example, the Beckman Coulter TL120.2 rotor with clear polycarbonate tubes (cat# 343778 from Beckman Coulter). See Note 4.
7. The supernatant should be assessed as described above in step 4.

3.2. Measuring Aggregation State by Gel Filtration Chromatography

Another assay for measuring aggregation is gel filtration chromatography. Gel-filtration chromatography separates molecules on the basis of the ability of a solute, such as a protein, to diffuse into and out of the stationary-phase relative to the rate of the bulk flow of the mobile-phase. In general, the rates of diffusion into the stationary-phase depend on the size and shape of the solute, the composition of the stationary-phase, and the temperature and buffer/solvent conditions. In practice, this means that small, rapidly diffusing molecules, such as the native protein (typically less than 100 kDa) move slowly through a gel-filtration column while larger particles, such as fibrils, are either partially or totally excluded from the stationary phase and hence move rapidly.

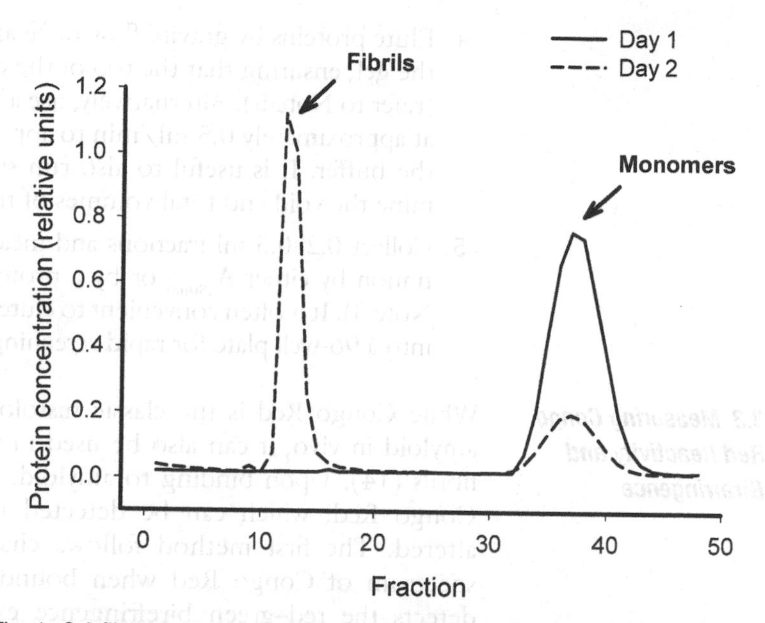

Fig. 1. Gel filtration chromatography to separate apoC-II monomers from fibrils as described in Subheading 3.2. The protein (0.5 ml) was loaded onto a 1 × 30 cm column packed in Sepharose CL-2B. The column was eluted in 100 mM sodium phosphate, pH 7.4 by gravity flow.

Amyloid-like fibrils tend to elute in the void volume of the most common resins and very large fibrils can be occluded from entering the column altogether – and this needs to be considered when using expensive prepacked columns, which can become irreversibly blocked. However, many smaller fibrils can be effectively separated on a column, and this provides a convenient means to measure the proportion of, for example, monomers from aggregates, which typically elute in the void volume (Fig. 1).

An inexpensive option is to pack a column with a low-cost resin, such as Sepharose CL-2B, which is a cross-linked agarose resin from GE Healthcare, and to elute the proteins by a peristaltic pump or gravity flow. This resin has a nominal fractionation range of approximately 70,000–4,000,000 Da for proteins and has a high flow rate capacity making it ideal for separating proteins that have a very large difference in size, such as fibrils from the unaggregated counterpart.

1. Pack a small column (e.g., 1 × 30 cm Econo-column from BioRad) with Sepharose CL-2B and equilibrate in buffer of choice. See Notes 5 and 6.
2. Equilibrate the column with several column volumes of buffer to ensure that the resin is fully settled (see Notes 7 and 8).
3. Load samples containing the protein carefully on the top of the column in no more than 0.5 ml for a 1 × 30 column (a good rule of thumb is to load no more than 0.02 column volumes of sample to reduce peak broadening).

4. Elute proteins by gravity flow once all the sample has entered the gel, ensuring that the top of the column does not dry out (refer to Note 5). Alternatively, use a peristaltic pump running at approximately 0.5 ml/min to apply the sample followed by the buffer. It is useful to also run sizing standards to determine the void and total volumes of the column (see Note 9).
5. Collect 0.2–0.5 ml fractions and measure the protein concentration by either A_{280nm} or by a protein detection kit (refer to Note 3). It is often convenient to elute 0.25 ml fractions directly into a 96-well plate for rapid screening in a plate reader.

3.3. Measuring Congo Red Reactivity and Birefringence

While Congo Red is the classic histological stain for detecting amyloid in vivo, it can also be used in vitro to test for amyloid fibrils (14). Upon binding to amyloid, the optical properties of Congo Red, which can be detected using two methods, are altered. The first method follows changes to the absorption spectrum of Congo Red when bound to fibrils. The second, detects the red–green birefringence exhibited by fibril-bound Congo Red under cross-polarized light.

3.3.1. Method 1: Congo Red Spectroscopic Assay

1. Prepare a fresh solution of 100 μM Congo Red in PBS (from the 1 mM stock in ethanol) and filter through a 0.22 μm syringe filter immediately prior to use to remove particulates and undissolved Congo Red.
2. Prepare test samples by mixing together the protein sample with the Congo Red solution. Ensure that the final Congo Red concentration is between 2 and 20 μM. Also the protein concentration (the concentration of monomeric protein) in the sample should generally not exceed 100 μM. See Note 10.
3. Incubate for 5 min at room temperature.
4. Prepare control samples. Control A is Congo Red alone at the same concentration and final buffer composition as that in step 2. This control defines the Congo Red reference spectrum. It is important to dilute Congo Red into the same buffer in which the protein has been prepared. Control B is the protein alone at the same concentration as the test samples, and also the same buffer composition (i.e., dilute the protein into PBS instead of the Congo Red solution). Control B enables a correction to be made to account for solution turbidity caused by protein aggregates.
5. Set a UV/Vis spectrophotometer to acquire absorbance spectra between 300 and 700 nm and zero the instrument against PBS. Measure the absorbance spectra of all test and control samples. At this point, a shoulder peak centered around 540 nm, which is indicative of amyloid fibrils, may be apparent in the test sample that is not visible in control sample A.

6. Subtract the spectrum of control B (protein alone) from the test spectrum to correct for turbidity. From the resultant spectrum, subtract the spectrum acquired for control A (Congo Red alone). A peak in the resulting final difference spectrum at 540 nm is generally considered indicative of amyloid fibrils. See Note 11.

3.3.2. Method 2: Congo Red Birefringence Assay

There are various ways in which protein aggregates can be prepared for the birefringent assay. All of them in essence condense Congo Red-labeled aggregates to masses large enough to be visualized by microscopy. The different options for preparation should be trialed for different proteins.

Preparation option A

1. Place 10–20 μl of aggregate suspension on a poly-lysine coated microscope slide and allow to air dry at room temperature or in a 37°C incubator.
2. Place ~50 μl of Congo Red stain solution onto the dried protein sample and incubate for a few minutes.
3. Blot away excess solution and allow to air dry once more.
4. If necessary, excess stain can be removed by washing the sample with a small volume of distilled water.

Preparation option B

1. Pellet 1 ml volume of aggregates by centrifugation and remove the supernatant.
2. Resuspend the aggregate pellet in Congo Red stain solution and incubate for 5 min.
3. Pellet the stained fibrils once more and resuspend in a small volume of distilled water. See Note 12.
4. Place the stained aggregate suspension on a poly-lysine-coated microscope slide and allow to air dry at room temperature or in a 37°C incubator.

Preparation option C (see Note 13)

1. Place protein solution containing aggregates in a centrifugation filtration device and concentrate proteins until large (> mm in size) clumps of protein appear.
2. Collect aggregates and resuspend in Congo Red stain solution and incubate for 5 min.
3. Pellet the stained aggregates once more and resuspend in a small volume of distilled water. Refer to Note 12.
4. Place the stained fibril suspension on a slide and cover with distilled water for imaging.

Imaging of birefringence

1. Examine the sample on a light microscope.
2. When the polarizers are aligned, Congophilic fibrils appear pink to red. When the polarizers are crossed at a 90° angle to one another, incident light is blocked and the background becomes dark. Under these conditions, amyloid fibrils bound to Congo Red appears as bright green areas in the stained sample. This birefringence is due to the ability of the bound dye to alter the polarization of the incident light and allow its transmission through the objective (see Notes 14 and 15).

3.4. Measuring Thioflavin T Reactivity

ThT is another diagnostic amyloid dye originally developed as a histological stain (15, 16). It was subsequently to be used in vitro to measure amyloid-like fibrils (17). In essence, the assay detects large spectral changes in the fluorescence properties of ThT when bound to amyloid fibrils, which can be measured with a fluorimeter. Hence, the assay works best in a comparative manner between the nonaggregated protein versus the aggregated counterpart.

Simple spectroscopic assay (see Note 16)

1. Dilute stock ThT to 10 μM in approximately 10 ml PBS.
2. Immediately measure the fluorescence intensity of 1 ml of this solution in a fluorescence cuvette using excitation at 445 nm (5 nm bandwidth) and emission at 482 nm (10 nm bandwidth), and average over several seconds.
3. Prepare a solution of the nonaggregated protein in a final concentration of 10 μM ThT. Generally, 10–100 μg/ml of protein is sufficient.
4. Measure the fluorescence intensity of this solution using parameters as above. This measurement serves as a negative control to reference against the (expected) increased fluorescence yield observed with the aggregated protein (in step 5). This control is important to account for ThT binding of the unaggregated form of the protein.
5. Prepare a solution of aggregated protein in a final concentration of 10 μM ThT using the same protein concentration as that in step 3, and measure fluorescence intensity as before. If the protein is amyloid, there should be a significant, and commonly very large, increase in fluorescence intensity relative to the unaggregated form of the protein (see Note 17).

3.5. Kinetics of Aggregation

Once it has been established that the protein has aggregated, it is often useful to determine the kinetics of aggregation and/or how environmental parameters might affect the kinetics. Doing so first requires an understanding of the oligomeric state of the native form of the protein in native-like buffers (e.g., it is monomeric, dimeric or some other assembly state in PBS), and preparation of the protein to ensure that the native state is free of large amyloid aggregates, which are the baseline, time zero reference sample.

There is no one-size fits all method to determine these properties of a protein. However, a good place to start is to measure the oligomeric size of the native state using gel filtration chromatography in a buffer that matches the native environment of the protein. Other more sophisticated and high resolution techniques for assessing oligomeric state include analytical ultracentrifugation and dynamic light scattering (these techniques are not covered in this chapter). Some proteins, particularly small proteins and peptides, can be stored denatured at high concentrations, such as in 5 M guanidine hydrochloride, and freshly refolded into the native state by directly diluting out of the denaturant. However, the efficiency of refolding and potential aggregation also needs to be assessed when using this approach.

Another important parameter is whether the native state can be stored frozen in aliquots, which enables more flexibility and convenience in performing the kinetic assays. A general framework for preparing the native protein for kinetic assays is provided here, which requires optimization using the guidelines described above for each particular protein. The volumes shown in the procedure here may need to be scaled depending on protein quantity required for downstream analysis, and type of kinetic assay to be performed.

1. Take a sample of the native form of the protein, and remove all preexisting aggregates by either high-speed centrifugation at 100,000×*g* as described above in Subheading 3.1 step 6, or by gel filtration chromatography as described above in Subheading 3.2.
2. Immediately place native protein on ice and set aside a small aliquot to measure the protein concentration.
3. Before measuring protein concentration, aliquot protein into single use volumes in microcentrifuge tubes and snap freeze in either liquid nitrogen, or alternatively, a dry ice-ethanol bath. See Note 18.
4. Place the aliquots at −80°C (for storage up to 6 months). See Note 19.
5. Using the protein set-aside, measure the concentration of the protein using a standard assay or by measuring the absorbance at 280 nm and molar extinction coefficient. Refer to Note 3. It is important to measure the protein concentration precisely for quantitative kinetic assays, since protein concentration strongly influences the aggregation rate of most amyloid proteins.
6. Thaw an aliquot of the protein and test whether the protein remains in the native oligomeric state by gel filtration

chromatography as described in Subheading 3.2 or by other high resolution techniques, such as analytical ultracentrifugation.

7. If the protein has formed aggregates by the freeze-thawing process, then either omit the freezing step for storage (note that the kinetics can still be performed, however, with less flexibility). Alternatively, optimize the buffer conditions to avoid the formation of aggregates by freeze-thawing.

Once the protein preparation is complete, it is possible to perform the amyloid fibrillization kinetic assays using one of the three following methods, the *reverse kinetic assay*, the *forward kinetic assay*, and the *continuous ThT assay*. See Note 20.

3.5.1. Method 1: Reverse Kinetic Assay

1. Devise the time points and time frame of the assay. Keep in mind that the assay will be run in reverse with the longest incubation times started first, and with aliquots progressively taken out of storage for decreasing incubation timepoints. At the time of the last aliquot (time point 0), all samples will be measured simultaneously for aggregation. Typically, fibrillization occurs on a time scale of minutes to days; however, this can vary widely depending on the protein and environmental conditions. A good starting point for 0.3 mg/ml protein is a 72 h time course, with time points heavily weighted over the first 24 h.
2. Take an aliquot of protein out of −80°C storage (or dilute out of denaturant).
3. Place in an incubator, PCR machine, or heat block, to regulate temperature.
4. Note down the time the tube was thawed.
5. Once the tube has warmed (several seconds to a minute), open the tube and pipette mineral oil (~200 μl) to cover the surface area of sample. This step is crucial to prevent buffer evaporation and condensation on the lid of the tube. However, the mineral oil can be omitted if using a PCR machine equipped with a heated lid option that prevents such condensation.
6. Close lid and continue incubating.
7. Repeat steps 2–6 for each decreasing time point.
8. For the last tube, thaw briefly at the incubation temperature and note the time of thawing.
9. Measure amyloid formation using the assays described (Subheadings 3.1–3.3) above for each aliquot. To remove the sample from the tube, carefully insert pipette tip below the mineral oil and aspirate slowly. It should be straightforward to pipette without drawing oil. See Note 21.

3.5.2. Method 2: Forward Kinetic Assay

1. This assay uses a continuous protein solution, or parallel array of tubes, which begin the assay simultaneously and are measured at time points of incrementing time as shown in Fig. 2.

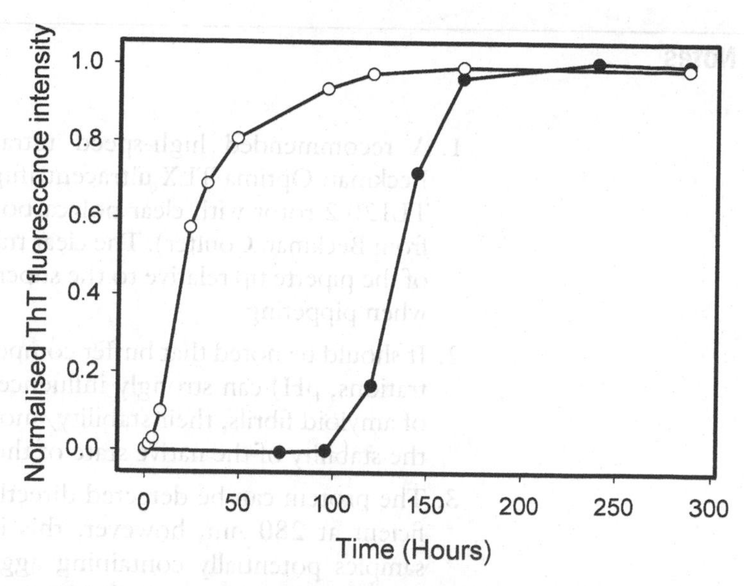

Fig. 2. Fibril formation time course for apolipoprotein C-II (open circles) and a peptide fragment derivative of apolipoprotein C-II, apoC-II_{56-76} (filled circles) monitored by the Method 2: Forward kinetic assay ThT assay (Subheading 3.5.2). The ThT fluorescence intensity has been standardized to the maximum ThT fluorescence intensity signal. However, the ThT fluorescence yield in the presence of apoC-II_{56-76} fibrils is approximately fivefold higher than that in the presence of full length apoC-II fibrils.

2. Thaw sufficient aliquots for assay (one aliquot per time point) and pool together. Alternatively, if the protein cannot be freeze-thawed, take freshly prepared protein as described above in steps 1–6 of Subheading 3.5.
3. Place protein in an incubator at a regulated temperature.
4. Overlay the protein with mineral oil as described above in step 5 of Subheading 3.5.1.
5. At the first timepoint, remove an aliquot from the tube and assay for amyloid as described in Subheadings 3.1–3.5.
6. Repeat step 5 for each timepoint until assay is complete.

3.5.3. Method 3: Continuous ThT Kinetic Assay

1. Prepare ThT stock and fluorimeter as described in Subheading 2.3. This protocol is most suited to use with a multi-well plate and plate reader (the following protocol will assume this setup). See Note 16.
2. Set plate reader fluorimeter to read fluorescence of the plate at defined intervals over a period appropriate for the time course of fibril formation. See Note 22.
3. Prepare protein for fibril formation in the presence of 10 μM ThT.
4. Place aliquots of this preparation in plate wells and seal with plate-sealing film.
5. Immediately place plate in plate reader and begin data collection.

4. Notes

1. A recommended high-speed ultracentrifuge setup is the Beckman Optima TLX ultracentrifuge and Beckman Coulter TL120.2 rotor with clear polycarbonate tubes (cat# 343778 from Beckman Coulter). The clear tubes facilitates the visibility of the pipette tip relative to the supernatant and protein pellet when pippeting.
2. It should be noted that buffer compositions (e.g., salt concentrations, pH) can strongly influence the formation and rates of amyloid fibrils, their stability/morphology, and conversely the stability of the native state of the protein.
3. The protein can be detected directly by the extinction coefficient at 280 nm, however, this is not recommended for samples potentially containing aggregates, which can have significant turbidity contributions at 280 nm. Hence, protein concentrations should be measured using standard protein detection assays, such as Lowry (18), bicinchoninic acid (19), or Coomassie Blue-based approaches (e.g., Bradford assay (20)). Such assays also provide more flexibility in protein volumes and concentrations and can be adapted to a plate reader format for large sample handling. Inexpensive kits are available from most bioscience vendors.
4. To readily identify the pellet, mark the outside face of the tube with a dot. It is important that the supernatant be removed by slow aspiration away from the pellet to avoid disturbing the pelleted protein.
5. If using gravity flow to elute the proteins, ensure that the column outlet contains a valve so that the flow can be regulated. Alternatively, the Econo-pac columns from BioRad are designed to stop flowing when the reservoir above the column has emptied, which is particularly useful for gravity flow applications. If using a peristaltic pump, place a flow adaptor to the top of the column to ensure that the sample is applied evenly to the resin.
6. Typically, the buffer should contain at least 150 mM salt to minimize nonspecific binding of the protein to the resin (e.g., 20 mM Tris, pH 7.8, 150 mM NaCl). Note also that long, thin columns provide the best resolution for gel filtration.
7. If using a peristaltic pump, adjust the flow adaptor to remove any gap that may have formed from further resin compaction and settling.
8. When packing and running columns be careful not to run the flow rate faster than the manufacturer's recommendation because the resin can compact over time leading to blocked

flow – often in the middle of a crucial run! Also, be sure to equilibrate the column with at least 1.5–2 column volumes for each run to ensure material binding nonspecifically is fully eluted.

9. Blue dextran 2000 (GE Healthcare) works well for labeling the void volume on a sepharose CL-2B column.
10. In the first instance, it is worth trying 2 μM Congo Red with as high a protein concentration as possible, but the ratio and final concentrations may need optimizing depending on the protein.
11. Other proteins that are not in an amyloid-state may also produce a similar result, and hence it is important to also measure birefringence, which is a more conclusive indicator of amyloid fibrils.
12. At this stage, it is possible to wash the stained fibrils by repeated pelleting and resuspension in distilled water, if required to remove excess stain.
13. This method is good for concentrating protein aggregates from large or dilute volumes.
14. The thickness of the sample can have a large effect on the observation of birefringence. Thus, a number of different samples of different spread and thickness should be examined.
15. Many things are capable of producing birefringence under cross-polarized light (e.g., hair and other natural fibers, dust, salt crystals). For this reason, false positives and misinterpretation can occur. Detection of strong *green* birefringence corresponding to areas stained pink/red under bright-field examination is generally considered positive for amyloid. Ideally, a known amyloid sample, such as Aβ should be examined in parallel. See Fig. 3 shows the birefringence for a protein prepared by Option C.
16. The ThT fluorescence assay can be easily adapted for multi-well plates and using a fluorescence plate reader. This allows the assay volume to be substantially reduced, thereby reducing the consumption of protein. To do this, modify the assay so that the final concentration of ThT is 10 μM. Hence, in a 96-well format, pipette together 200 μl 12.5 μM ThT in PBS with 50 μl protein solution.
17. The fluorescence yield of ThT can be quite different in the presence of different types of amyloid fibrils. Thus, it may be necessary to vary the protein concentration used in the assay to optimize signal and gain statistically significant results.
18. If using dry ice ethanol bath, be careful not to submerse cap of microcentrifuge tube under the ethanol which can lead to ethanol leaking into the tubes. Also use an alcohol-resistant marker on the tubes to avoid labeling washing off.

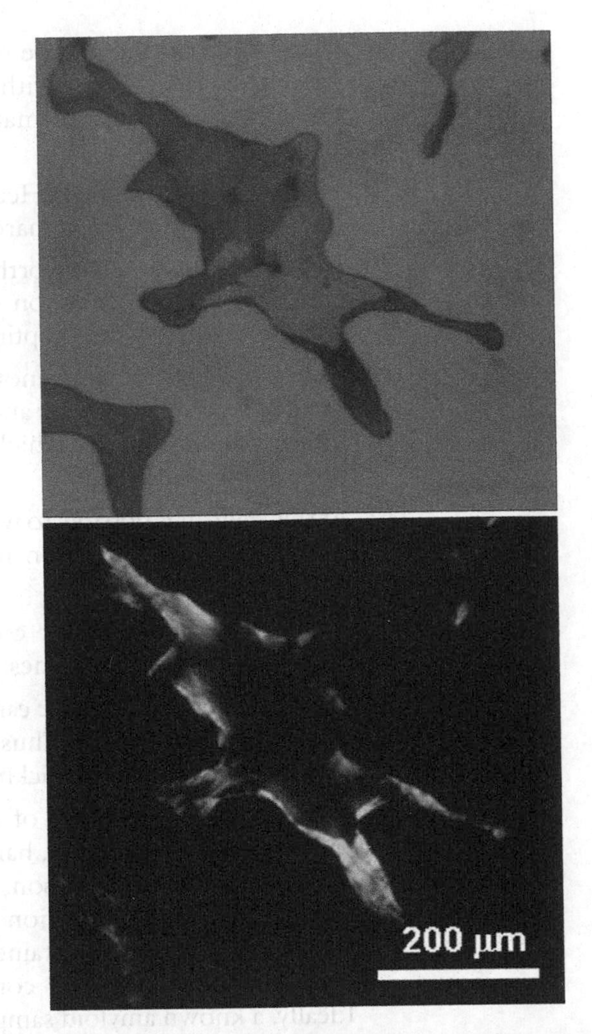

Fig. 3. Red–green Congo Red birefringence of amyloid fibrils formed by apolipoprotein C-II. The protein aggregates were prepared according to Option C in Subheading 3.3.2, step 3. The upper panel is the brightfield view of the Congo Red stained aggregates. The lower panel shows the green birefringence under cross-polarized light.

19. Do not store at −20°C, which commonly facilitates protein aggregation. It is important to use the snap-freeze procedure that, in general, induces less protein aggregation.

20. The *reverse kinetic assay* can only be performed if the protein can be freeze-thawed as described above without inducing aggregation or if the protein can be diluted directly out of a denatured stock without causing spontaneous aggregation. The forward kinetic assay has the advantage of working with proteins that cannot be freeze-thawed. However, drawbacks of this assay are that there is the potential for baseline drifts over long periods (especially for measuring ThT reactivity)

and that it is more time consuming. The continuous ThT assay offers the huge advantage in that it is simple, and easy to scale up, however, has a major potential drawback in that the ThT can change fibrillization rates, and hence this effect needs to be considered when using this assay.

21. A small amount of oil carryover from the outside of the tip usually does not interfere with the assays for amyloid formation. The most convenient assay for kinetic analysis is the ThT assay in plate reader format. A multichannel pipettor is very handy for large-scale kinetic analysis using the ThT assay in the plate reader format.

22. It is important that the fibril suspension is adequately mixed before each fluorescence measurement, as fibrils can become flocculent and sediment quickly to the bottom of the plate well. If your plate reader has a "mix" functionality, program the instrument to thoroughly mix the plate before each data point collection.

References

1. Virchow R. (1854) Ueber eine im gehirn und ruckenmark des menschen aufgefunde substanz mit der chemishen reaction der cellulose. *Virchows Arch Path Anat* 6, 135–138.
2. Divry P., and Florkin M. (1927) Sur les proprietes optiques de l'amyloide.
3. *Comptes Rendus de la Societe de Biologie* **97**, 1808–1810.
4. Missmahl H. P., and Hartwig M. (1953) Polarisationsoptische untersuchungen an der amyloidsubstanz. *Virchows Archiv* **324**, 489–508.
5. Westermark P., Benson M. D., Buxbaum J. N., Cohen A. S., Frangione B., Ikeda S.-I., Masters C. L., Merlini G., Saraiva M. J., and Sipe J. D. (2005) Amyloid: Toward terminology clarification report from the nomenclature committee of the international society of amyloidosis. *Amyloid* **12**, 1–4.
6. Cohen A. S., and Calkins E. (1959) Electron microscopic observations on a fibrous component in amyloid of diverse origins. *Nature* **183**, 1202–1203.
7. Chiti F., and Dobson C. M. (2006) Protein misfolding, functional amyloid, and human disease. *Ann. Rev. Biochem.* **75**, 333–366.
8. Sipe J. D., and Cohen A. S. (2000) Review: History of the amyloid fibril. *J. Struct. Biol.* **130**, 88–98.
9. Serpell L. C. (2000) Alzheimer's amyloid fibrils: Structure and assembly. *Biochim. Biophys. Acta* **1502**, 16–30.
10. Sunde M., and Blake C. (1997) The structure of amyloid fibrils by electron microscopy and x-ray diffraction. *Adv. Protein Chem.* **50**, 123–159.
11. Booth D. R., Sunde M., Bellotti V., Robinson C. V., Hutchinson W. L., Fraser P. E., Hawkins P. N., Dobson C. M., Radford S. E., Blake C. C., and Pepys M. B. (1997) Instability, unfolding and aggregation of human lysozyme variants underlying amyloid fibrillogenesis. *Nature* **385**, 787–793.
12. Fandrich M., Fletcher M. A., and Dobson C. M. (2001) Amyloid fibrils from muscle myoglobin. *Nature* **410**, 165–166.
13. Chapman M. R., Robinson L. S., Pinkner J. S., Roth R., Heuser J., Hammar M., Normark S., and Hultgren S. J. (2002) Role of Escherichia coli curli operons in directing amyloid fiber formation. *Science* **295**, 851–855.
14. Hatters D. M., Zhong N., Rutenber E., and Weisgraber K. H. (2006) Amino-terminal domain stability mediates apolipoprotein E aggregation into neurotoxic fibrils. *J. Mol. Biol.* **361**, 932–944.
15. Klunk W. E., Jacob R. F., and Mason R. P. (1999) Quantifying amyloid β-peptide (Aβ) aggregation using the congo red-Aβ (CR-Aβ) spectrophotometric assay. *Anal Biochem* **266**, 66–76.
16. Vassar P. S., and Culling C. F. (1959) Fluorescent stains, with special reference to amyloid and connective tissues. *Arch. Pathol.* **68**, 487–498.

17. Kelenyi G. (1967) On the histochemistry of azo group-free thiazole dyes. *J Histochem. Cytochem.* **15**, 172–180.
18. LeVine H., 3rd. (1993) Thioflavine T interaction with synthetic alzheimer's disease β-amyloid peptides: Detection of amyloid aggregation in solution. *Protein Sci.* **2**, 404–410.
19. Lowry O. H., Rosebrough N. J., Farr A. L., and Randall R. J. (1951) Protein measurement with the Folin phenol reagent. *J. Biol. Chem.* **193**, 265–275.
20. Smith P. K., Krohn R. I., Hermanson G. T., Mallia A. K., Gartner F. H., Provenzano M. D., Fujimoto E. K., Goeke N. M., Olson B. J., and Klenk D. C. (1985) Measurement of protein using bicinchoninic acid. *Anal Biochem* **150**, 76–85.
21. True H. L., and Lindquist S. L. (2000) A yeast prion provides a mechanism for genetic variation and phenotypic diversity. *Nature* **407**, 477–483.
22. Hatters D. M., MacPhee C. E., Lawrence L. J., Sawyer W. H., and Howlett G. J. (2000) Human apolipoprotein C-II forms twisted amyloid ribbons and closed loops. *Biochemistry* **39**, 8276–8283.
23. Bradford M. M. (1976) A rapid and sensitive method for the quantitation of microgram quantities of protein utilizing the principle of protein-dye binding. *Anal Biochem* **72**, 248–254.

Chapter 9

Probing Protein Aggregation with Quartz Crystal Microbalances

Tuomas P. J. Knowles, Glyn L. Devlin, Christopher M. Dobson, and Mark E. Welland

Abstract

The supra-molecular self-assembly of peptides and proteins is a process which underlies a range of normal and aberrant biological pathways in nature, but one which remains challenging to monitor in a quantitative way. We discuss the experimental details of an approach to this problem which involves the direct measurement in vitro of mass changes of the aggregates as new molecules attach to them. The required mass sensitivity can be achieved by the use of a quartz crystal transducer-based microbalance. The technique should be broadly applicable to the study of protein aggregation, as well as to the identification and characterisation of inhibitors and modulators of this process.

Key words: Quartz crystal microbalance, Biosensors, Kinetics

1. Introduction

The folding of soluble proteins into their native states is in general an essential requirement for their biological activity. Much research has focussed on understanding the mechanisms by which polypeptide chain acquires these active conformations, and more recently, the processes which can lure proteins away from such states towards the formation of intractable aggregates known as amyloid fibrils (1, 2). The aberrant assembly in vivo of proteins and peptides into amyloid fibrils or their precursors is a process which is associated with a wide range of clinical disorders (3, 4), including Alzheimer's and Parkinson's diseases. Nature has, however, also found functional applications for the amyloid fold in for

Andrew F. Hill et al. (eds.), *Protein Folding, Misfolding, and Disease: Methods and Protocols*, Methods in Molecular Biology, vol. 752, DOI 10.1007/978-1-60327-223-0_9, © Springer Science+Business Media, LLC 2011

instance the catalysis of at least one polymerisation reaction (5), the formation of fibrous materials, e.g. for bacterial coatings (6) and the non-genetic transmission of information in certain fungi (7–9). The importance of these varied roles of amyloid fibrils in biology highlights the need for the elucidation of the underlying mechanisms by which these species form and the characterisation of the kinetic parameters governing the rates at which fibril assembly occurs under given conditions.

Currently, one of the most important sources of information on the mechanistic determinants that underlie protein aggregation is from kinetic measurements. Due to the complexity of the process, however, and the multiple on and off pathway steps that a protein system can access during the aggregation reaction, such kinetic studies remain challenging. In particular, quantitative measurements of the rates at which fibrils grow through the progressive addition of protein molecules from the solution have proved to be difficult to achieve. Solution-based optical methods are often employed to make these measurements; they are, however, limited by various factors. Light scattering measurements for example are sensitive to processes extraneous to fibril elongation such as non-specific amorphous aggregation and fibril–fibril association, while the inference in fibril growth kinetics from dye-binding measurements (using amyloidophilic dyes such as thioflavin T and Congo red) requires a prior knowledge of the stoichiometery and specificity of the binding of these dyes to the fibrils of interest. We discuss here the experimental details of an alternative approach for monitoring protein aggregation which is based on measuring directly as a function of time the changes in hydrodynamic mass resulting from fibril growth (10–15).

2. Materials

1. Bovine insulin (Sigma).
2. 1N HCl (Sigma).
3. Probe sonicator.
4. Mercapto-poly(ethylene glycol) (mercapto-PEG) $CH_3O(CH_2CH_2O)_6CH_2SH$ (Polypure, Norway).
5. QSX 301-Standard gold-coated quartz crystals from Q-Sense (Q-Sense AB, Västra, Frölunda, Sweden) with a frequency/mass sensitivity coefficient of 17.7 and fundamental frequency of 4.95 MHz.
6. D300 QCM flow cell.

3. Methods

3.1. Sensor Functionalisation

In order to direct the deposition of protein onto the QCM sensor crystal/solution interface, fragments of pre-formed fibrils were attached to the sensor surface. These fragments can act as seeds and elongate in the presence of precursor protein in solution. As the natural sequence of insulin comprises six sulfhydryl-containing cysteine residues, the fixation of the seed fibril fragments can be mediated through strong gold–sulphur interactions (see Note 1). For sequences without cysteine residues, additional steps such as those outlined in refs. 16–18 are required (Fig. 1).

1. Insulin amyloid fibrils were prepared by incubating bovine insulin (Sigma Aldrich) at 10 mg/ml in an aqueous solution adjusted to pH 2.0 with hydrochloric acid at 60°C for 12 h and subsequently for 7 days at room temperature. Seed fragments were then prepared by diluting this stock insulin fibril solution 1:500 into 10 mM HCl and exposing to an ultrasound treatment for 10 min using a probe sonicator. After this procedure, a homogeneous dispersion of seed fibril fragments of a length of approximately 100 nm was produced (see Note 2).

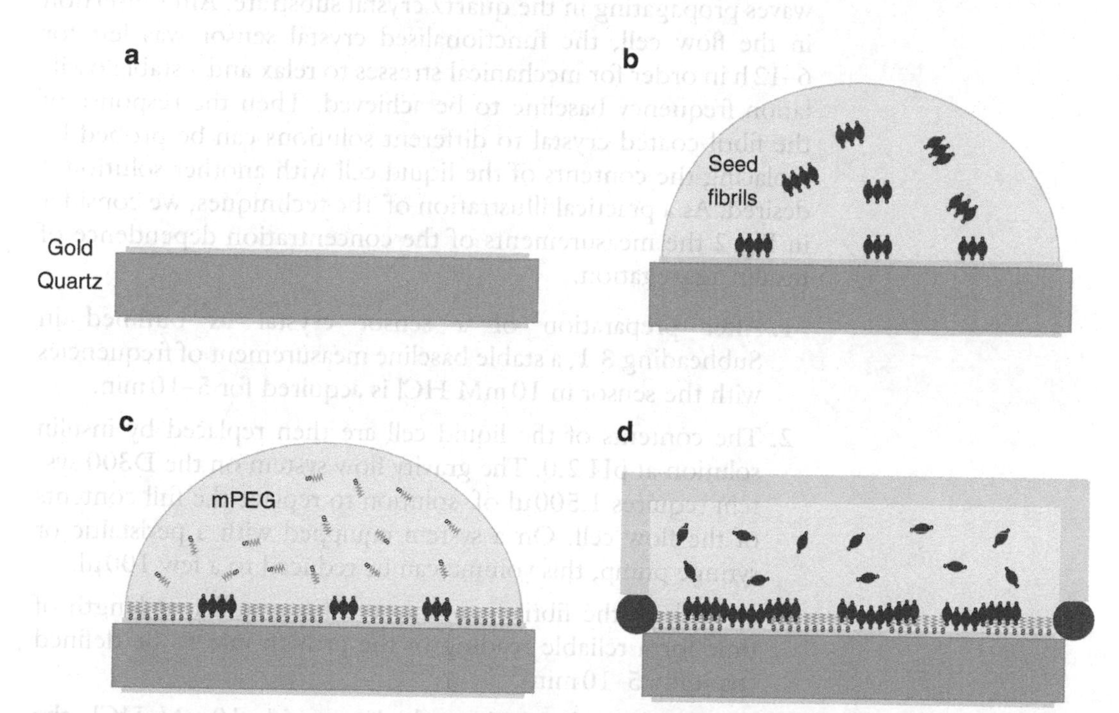

Fig. 1. QCM measurements of amyloid growth. The sensor is composed of a quartz crystal which includes gold electrodes (**a**). The top electrode is then functionalised with seed fibril fragments (**b**), the unexposed area passivated using mercapto-PEG (**c**), and the sensor inserted into a flow cell where the frequency response of the surface shear waves can be measured in the presence of monomeric protein ((**d**), *blue ovals*) or other compounds that affect the growth rate.

2. In a second step 150 μl of the suspension containing short fibril fragments was incubated on the gold surface of the sensor crystal for 60 min in an atmosphere of 100% humidity to allow for the adsorption of fibril segments on the surface without evaporation of the solvent (see Notes 3 and 4).
3. The surface area not covered with adsorbed fibrils was then passivated with an inert self-assembling polymer layer to prevent non-specific protein adsorption. For this purpose, a 0.05% mercapto-PEG solution in 10 mM HCl was used to wash the top sensor side of the crystal three times, and then 150 μl of the mercapto-PEG solution was loaded and left on the sensor for 60 min (see Note 4).
4. After such treatment, the sensor crystal was washed with 10 mM HCl to remove the free mercapto-PEG. The sensor was then immediately inserted into the flow cell and washed with 2 ml of 10 mM HCl (see Note 4).

3.2. Kinetic Measurements

We measure the growth kinetics of amyloid fibrils through the increase in hydrodynamic mass resulting from the attachment of new molecules to the fibrillar aggregates on the surface. Real-time changes in mass on the order of nanograms can accurately be measured through the shift in the resonant frequency of shear waves propagating in the quartz crystal substrate. After insertion in the flow cell, the functionalised crystal sensor was left for 6–12 h in order for mechanical stresses to relax and a stable oscillation frequency baseline to be achieved. Then the response of the fibril-coated crystal to different solutions can be probed by replacing the contents of the liquid cell with another solution if desired. As a practical illustration of the techniques, we consider in Fig. 2 the measurements of the concentration dependence of insulin aggregation.

1. After preparation of a sensor crystal as outlined in Subheading 3.1, a stable baseline measurement of frequencies with the sensor in 10 mM HCl is acquired for 5–10 min.
2. The contents of the liquid cell are then replaced by insulin solution at pH 2.0. The gravity flow system on the D300 system requires 1,500 μl of solution to replace the full contents of the flow cell. On a system equipped with a peristaltic or syringe pump, this volume can be reduced to a few 100 μl.
3. Growth of the fibrils is monitored for a sufficient length of time for a reliable reading of the growth rate to be defined (typically 5–10 min).
4. The insulin solution is washed out with 10 mM HCl, the growth stops in the absence of insulin monomers in solution, and after a stable baseline is achieved, the growth of the same ensemble of fibrils can be probed many times under different conditions.

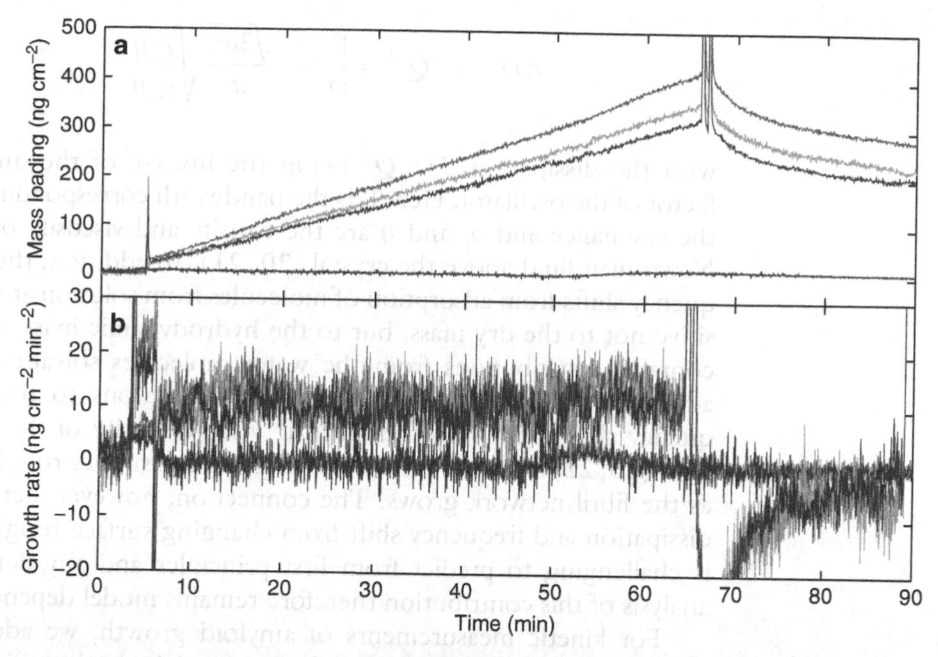

Fig. 2. Probing amyloid growth through measurements of hydrodynamic mass changes. (**a**) A sensor prepared as described in Subheading 3.1 is initially in contact with 10 mM HCl (0–5 min) and then exposed to 2 mg/ml insulin solution at 25°C starting (5–65) for 1 h resulting in growth of the fibrils on the sensor which can be monitored through the linear increase in mass. Data from (11) are superimposed which shows the absence of mass loading in the presence of insulin but absence of seed fibrils, demonstrating that the mass changes measure are specifically of the fibrils on the surface. In (**b**) the derivative of the data in (**a**) is shown giving the growth rate. The data in (**b**) were smoothed with a rectangular window transform of width 3 min. The frequency shifts in (**a**) and (**b**) (overtones from top to bottom: $n=3$, $n=5$, $n=7$) were converted to equivalent hydrodynamic masses using Eq. 1.

3.3. Analysis of the Results

For a thin and rigid adsorbed over-layer on a crystal oscillating in shear mode, the decrease in the resonant frequency in response to an increase in the surface mass loading is described by the Sauerbrey equation (19):

$$\Delta f_n = -n\frac{2f_n^2}{\sqrt{\rho_Q \mu}}\frac{\Delta m}{A} \tag{1}$$

where $\Delta m/A$ is the mass loading on the surface of the crystal, f_n is the resonant frequency of the unloaded crystal, n is the overtone number, and ρ_Q and μ are the density and shear modulus of Quartz.

When one side of the oscillating crystal is immersed in liquid, the free resonant frequency has an additional shift due to viscous damping from the fluid:

$$\Delta f \quad = \quad -2\pi\frac{\omega_0^{3/2}}{\pi}\sqrt{\frac{\rho_L \eta}{2\rho_Q \mu}} \tag{2}$$

$$\Delta D = Q^{-1} = \frac{\Gamma}{\omega} = -\frac{\sqrt{2\omega_0}}{\pi}\sqrt{\frac{\rho_L \eta}{\rho_Q \mu}} \quad (3)$$

with the dissipation $D = Q^{-}$ being the inverse of the quality factor of the oscillator. Here Γ is the bandwidth corresponding to the resonance and ρ_L and η are the density and viscosity of the Newtonian fluid above the crystal (20, 21). In addition, the frequency shifts from adsorption of molecules from solution are sensitive not to the dry mass, but to the hydrodynamic mass which contains contributions from the water molecules solvating the adsorbed molecules. More complex contributions to the frequency shift stem from the intrinsic visco-elasticity of the fibril layer (13, 22) as well as from the changes in the surface roughness as the fibril network grows. The connection, however, between dissipation and frequency shift from changing surface roughness is challenging to predict from first principles and any detailed analysis of this contribution therefore remains model dependent.

For kinetic measurements of amyloid growth, we adopt a model-free approach and consider nominal hydrodynamic frequency shifts computed from Eq. 1 (Figs. 2 and 3); these frequency shifts have empirically been shown to correlate well with the increase in length of fibrils resulting from their growth (11). If desired, atomic force microscopy measurements of fibrils before and after growth can be used to refine the calibration of frequency shifts in terms of absolute length increase of fibrils (11). For many measurements such as those discussed here of the effect of intrinsic or extrinsic factors on the rate of growth of fibrils, only relative rate information is required and an absolute calibration is not necessary. In all cases, sequential measurements conserve the number of growth sites on the sensor, and therefore the relative changes in the elongation rates of fibrils in response to changes in the environmental conditions can be determined with a high level of precision. This conservation in fibril number is in contrast to experiments in solution, where spontaneous fracturing of fibrils leads to an increase in the number of growing ends over time.

In practice, typically several overtones (for instance $n = 3,5,7$) are measured simultaneously, and the fundamental mode $n = 1$ is neglected as this is generally more prone to instabilities originating from interface effects due to the less efficient energy confinement for lower modes (24). Growth rates computed from the frequency shifts for the different overtones using Eq. 1 typically differ less than 30% for the lower overtones.

The data that can be obtained through this approach are illustrated in Figs. 2 and 3. Figure 2a shows the changes in the resonant frequency of the sensor system during fibril growth; the data have been converted to effective hydrodynamic mass variations using Eq. 1; the time derivative of the mass loading provides a

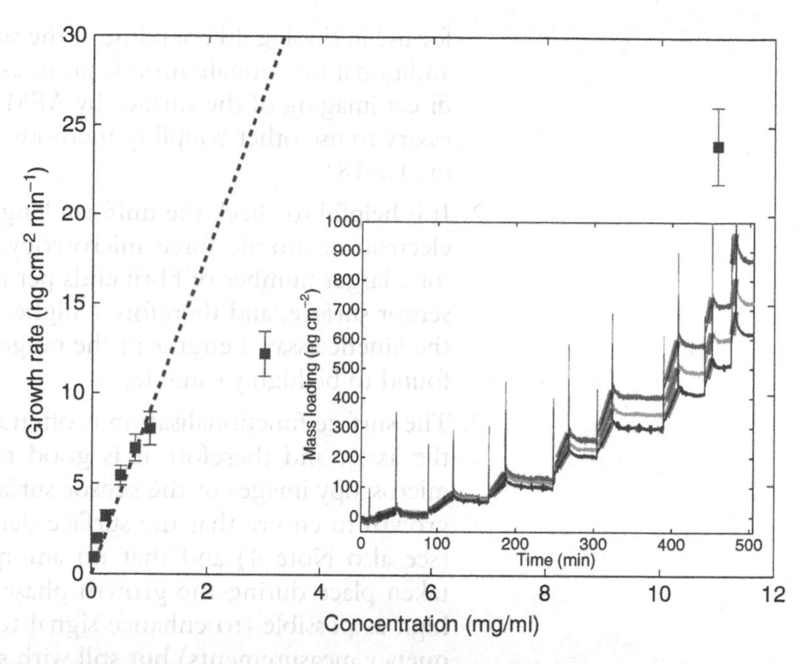

Fig. 3. Concentration dependence of amyloid fibril growth. Data from ref. (11). The *inset* shows raw QCM frequency shifts for overtones $n=3$, $n=5$, and $n=7$ converted to equivalent hydrodynamic masses with Eq. 9.1, and the main graph shows the average (*square*) and standard deviation (*errorbar*) of the data in the *inset* for the different concentrations. The initial linear trend is indicative of a second-order reaction between the fibril end and free monomers in solution, and the saturation behaviour observed at higher concentration suggests that the formation or disintegration of reaction intermediates becomes the rate-limiting step in this regime. This crossover point between these two types of behaviour occurs at a similar concentration as that measured for the growth of insulin spherulites (23). The frequency shifts in a and b (overtones from top to bottom: $n=3$, $n=5$, $n=7$) were converted to equivalent hydrodynamic masses using Eq. 1.

measure of the instantaneous fibril elongation rate as shown in Fig. 2b. The intrinsic or extrinsic aggregation conditions can be changed readily and the growth of fibrils probed repeatedly under different conditions. This idea is illustrated in Fig. 3 which shows the determination of the concentration dependence of amyloid fibril elongation for the protein insulin. The data reveal a linear dependence at low concentration of soluble proteins, and a transition to a less than linear dependence at higher concentrations, thereby suggesting the presence of intermediate states between the soluble and fibrillar forms of the protein.

4. Notes

1. Even in cases where the polypeptide sequence in question naturally contains cysteine residues, these may not be present at the surface of the fibrils and might therefore not be accessible

for use in thiol–gold coupling. If the surface coverage without additional functionalisation is poor, as shown for example by direct imaging of the surface by AFM (see Note 3), it is necessary to use other coupling methods such as those described in (16–18).

2. It is helpful to check the uniform length of the seed fibrils by electron or atomic force microscopy. Shorter lengths allow for a larger number of fibril ends per unit surface area on the sensor surface, and therefore a higher signal to noise ratio in the kinetic assay. Lengths in the range 50–200 nm have been found to be highly suitable.
3. The surface functionalisation is often the most critical step in the assay, and therefore it is good to acquire atomic force microscopy images of the sensor surface before and after the growth to ensure that the surface density is sufficiently high (see also Note 4) and that no amorphous aggregation has taken place during the growth phase. An ideal density is as high as possible (to enhance signal to noise ratio in the frequency measurements) but still with some space for fibrils to grow. We have observed that densities between 20 and 200 fibrils μm^{-2} typically yield good results, but a few trials with varying incubation times and/or seed fibril concentration are often required to achieve optimal results.
4. Drying of the sensor should be avoided. We have observed that once dried, the ability of the seed fibrils to promote further growth is diminished; this fact could be due to structural changes upon drying, which could affect the seeding ability of the fibrils. Therefore, AFM imaging of the sensor should be performed in liquid, or alternatively, samples for AFM in air can be prepared on gold substrates in parallel with the sensor functionalisation and using an identical protocol. After the washing step, the AFM substrates can be dried, and AFM images can be acquired in air. The preparation of flat gold substrates for AFM imaging is beyond the scope of the present text; we have obtained good results with evaporation of chromium (2–20 nm) and then gold (10–100 nm) onto a freshly cleaved mica surface.

References

1. F. Chiti, C. M. Dobson, *Protein misfolding, functional amyloid, and human disease*, Annu Rev Biochem **75**, 333 (2006).
2. C. M. Dobson, *Protein folding and misfolding*, Nature **426**, 884 (2003).
3. M. B. Pepys, *Pathogenesis, diagnosis and treatment of systemic amyloidosis*, Phil. Trans. R. Soc. Lond. B **356**, 203 (2001).
4. A. Aguzzi, *Understanding the diversity of prions*, Nat Cell Biol **6**, 290 (2004).
5. D. M. Fowler, A. V. Koulov, W. E. Balch, J. W. Kelly, *Functional amyloid–from bacteria to humans*, Trends Biochem Sci **32**, 217 (2007).
6. M. R. Chapman, L. S. Robinson, J. S. Pinkner, R. Roth, J. Heuser, M. Hammar, S. Normark,

S. J. Hultgren, *Role of Escherichia coli curli operons in directing amyloid fiber formation*, Science **295**, 851 (2002).

7. R. B. Wickner, *[URE3] as an altered URE2 protein: evidence for a prion analog in Saccharomyces cerevisiae.*, Science **264**, 566 (1994).
8. R. Krishnan, S. L. Lindquist, *Structural insights into a yeast prion illuminate nucleation and strain diversity*, Nature **435**, 765 (2005).
9. M. Tanaka, S. R. Collins, B. H. Toyama, J. S. Weissman, *The physical basis of how prion conformations determine strain phenotypes*, Nature **442**, 585 (2006).
10. C. M. Dobson, T. P. J. Knowles, M. E. Welland, UK patent application 0609382.7 (2006).
11. T. P. J. Knowles, W. Shu, G. L. Devlin, S. Meehan, S. Auer, C. M. Dobson, M. E. Welland, *Kinetics and thermodynamics of amyloid formation from direct measurements of fluctuations in fibril mass*, Proc Natl Acad Sci U S A **104**, 10016 (2007).
12. D. A. White, A. K. Buell, C. M. Dobson, M. E. Welland, T. P. J. Knowles, *Biosensor-based label-free assays of amyloid growth.*, FEBS Lett **583**, 2587 (2009).
13. M. B. Hovgaard, M. Dong, D. E. Otzen, F. Besenbacher, *Quartz crystal microbalance studies of multilayer glucagon fibrillation at the solid-liquid interface.*, Biophys J **93**, 2162 (2007).
14. H. Okuno, K. Mori, T. Okada, Y. Yokoyama, H. Suzuki, *Development of aggregation inhibitors for amyloid-beta peptides and their evaluation by quartz-crystal microbalance.*, Chem Biol Drug Des **69**, 356 (2007).
15. J. A. Kotarek, K. C. Johnson, M. A. Moss, *Quartz crystal microbalance analysis of growth kinetics for aggregation intermediates of the amyloid-beta protein.*, Anal Biochem (2008).
16. K. Hasegawa, K. Ono, M. Yamada, H. Naiki, *Kinetic modeling and determination of reaction constants of Alzheimer's beta-amyloid fibril extension and dissociation using surface plasmon resonance*, Biochemistry **41**, 13489 (2002).
17. A. K. Buell, G. G. Tartaglia, N. R. Birkett, C. A. Waudby, M. Vendruscolo, X. Salvatella, M. E. Welland, C. M. Dobson, T. P. J. Knowles, *Position-dependent electrostatic protection against protein aggregation.*, Chembiochem **10**, 1309 (2009).
18. D. A. White, A. K. Buell, T. P. J. Knowles, M. E. Welland, C. M. Dobson, *Protein aggregation in crowded environments.*, J Am Chem Soc **132**, 5170 (2010).
19. G. Sauerbrey, *Verwendung von Schwingquarzen zur Wägung dünner Schichten und zur Mikrowägung*, Z. Phys **155**, 206 (1959).
20. T. Nomura, O. M, *Frequency shifts of piezoelectric quartz crystals immersed in organic liquids*, Anal. Chim. Acta **142**, 281 (1981).
21. J. J. Kanazawa, J. G. Gordon, *Frequency of a Quartz Microbalance in Contact with Liquid*, Anal. Chem **57**, 1771 (1985).
22. M. V. Voinova, M. Rodahl, M. Jonson, K. B, *Viscoelastic Acoustic Responses of Layered Polymer Films at Fluid-Solid Interfaces: Continuum Mechanics Approach*, Physica Scripta **59**, 391 (1997).
23. S. S. Rogers, M. R. H. Krebs, E. H. C. Bromley, E. van der Linden, A. M. Donald, *Optical microscopy of growing insulin amyloid spherulites on surfaces in vitro*, Biophys J **90**, 1043 (2006).
24. I. Reviakine, A. N. Morozov, F. F. Rosetti, *Effects of finite crystal size in the quartz crystal microbalance with dissipation measurement system: Implications for data analysis*, J. Appl. Phys **95**, 7712 (2004).

Chapter 10

Dried and Hydrated X-Ray Scattering Analysis of Amyloid Fibrils

Sally L. Gras and Adam M. Squires

Abstract

Wide angle X-ray scattering is a key technique for the analysis of amyloid fibrils that can be used to confirm the presence of a characteristic cross-beta fibril structure and to characterise the arrangement of beta-strands and beta-sheets within this fibril core. Further structural insight can be obtained by the comparison of X-ray scattering data obtained for dried and hydrated fibril samples. We describe simple techniques for the preparation of dried and hydrated fibril samples for X-ray analysis and the subsequent analysis of X-ray scattering patterns using custom built and readily available software.

Key words: WAXS, X-ray, Fibre diffraction, Alignment, Hydrated, Software, Analysis

1. Introduction

In this article, we describe how to prepare hydrated and dried amyloid fibril samples for X-ray scattering analysis, how to obtain X-ray scattering patterns and how to analyse and interpret these patterns. This type of X-ray experiment is best described as "wide-angle X-ray scattering" ("WAXS") or "X-ray diffraction". Both terms are appropriate for fibrils; the former term generally refers to methods that probe the 0.2–2 nm length scale, the latter describes X-ray scattering experiments that give sharp reflections from features that are regularly repeated in space. Fibre diffraction is another term used to describe X-ray analysis, specifically of aligned fibrillar samples.

X-ray experiments can be used to confirm the presence of beta-sheet secondary structure and to determine the spacing of structural features within the fibril; namely, the beta-strands that stack up and down the length of the beta-sheets with a regular

Andrew F. Hill et al. (eds.), *Protein Folding, Misfolding, and Disease: Methods and Protocols*, Methods in Molecular Biology, vol. 752, DOI 10.1007/978-1-60327-223-0_10,

spacing of approximately 4.7 Å, and the beta-sheets that stack across the fibril, with a spacing of around 8–12 Å (1). Anisotropic X-ray patterns generated from aligned amyloid fibril samples can also be used to confirm the perpendicular arrangement of beta-strands and beta-sheets within the fibril core; this is known as a cross-beta confirmation and it is a key criterion that distinguishes amyloid fibrils from other fibrous and beta-sheet rich structures. The WAXS analysis of amyloid fibrils is, therefore, an important characterisation tool for confirming the structural traits of amyloid fibrils and furthering our understanding of fibril structure.

A number of other X-ray diffraction techniques can be used to obtain complementary structural information for amyloid fibril samples. Fibrils formed from short peptide sequences can be crystallised and the subsequent X-ray data used to build atomic structural models (2). Small angle X-ray scattering (SAXS) has also been used to observe the width of fibrils subjected to shear flow (3) and to characterise the structure of early oligomeric species during fibril assembly (4).

1.1. Aligned and Unaligned Samples

Sample alignment is a key determinant of the appearance of X-ray scattering patterns. Most liquids and powders are not orientated and generate an "isotropic" pattern which contains a series of concentric rings with no overall orientation. In contrast, aligned samples give rise to oriented scattering patterns. If amyloid fibrils are prepared in a dried stalk where the fibrils are aligned in the direction of the long axis of the stalk, the X-ray intensity appears in arcs; either to the left and right, or above and below the beam centre, depending on whether the reflection arises from periodicity parallel or perpendicular to the fibril axis. If the stalk of amyloid fibrils is positioned vertically in the X-ray beam, the inter-strand reflection at 4.7 Å is "axial" and so shows arcs above and below the main beam while the inter-sheet reflection at around 8–12 Å is "equatorial" and so shows arcs to the left and right. Similar patterns are observed for samples of amyloid fibrils aligned by other methods, such as from liquid flow in a capillary tube or in a flow cell (5–7). For highly oriented samples, these arcs can become still more aligned and the diffraction pattern appears as spots. In such cases, more advanced fibre diffraction analysis, outside the scope of this paper, may be carried out (8).

1.2. Considerations of Sample Hydration and Dehydration

Sample hydration is a further consideration for the analysis of structure by X-ray scattering. The processes used to induce sample alignment usually involve dehydration, which can potentially introduce structural artefacts. In 2004, Kishimoto and colleagues observed a clear difference in the structure of fibrils formed from the Sup35 yeast prion protein in the hydrated and dehydrated states (9). The reflection at 9 Å corresponding to the spacing between beta-sheets was present in X-ray patterns collected for dried fibrils but was absent in hydrated samples. We have since found that fibrils formed from two model proteins, $TTR_{105-115}$ and hen egg white lysozyme, give rise to identical reflections in the hydrated and dried states (5). It is, therefore, important to compare the X-ray scattering patterns obtained from an amyloid fibril sample in both dried and hydrated form.

1.3. Overview of This Article

We describe four key stages in the process of preparing and analysing amyloid fibrils by X-ray diffraction. These are:

(a) The preparation of dried stalks from amyloid fibril samples. This method can generate well-aligned samples.

(b) The preparation of hydrated samples in thin-walled glass capillary tubes. This method can give rise to partially aligned samples (although the process does involve a certain amount of chance).

(c) The experimental setup and collection of X-ray scattering patterns. We describe methods for loading samples in line with the X-ray beam and typical experimental parameters for obtaining an X-ray scattering pattern.

(d) The analysis of X-ray scattering patterns. We describe methods for processing 2-D X-ray data (i.e. scattering pattern images) to give 1-D data (i.e. *x–y* plots of intensity as a function of radius or d-spacing), to determine the positions and orientations of X-ray reflections. We also describe methods to subtract the background intensity contributed by water (or other solvents) for data obtained from hydrated samples.

A number of other methods exist for preparing amyloid fibril samples for X-ray diffraction analysis that is not reviewed here. These include drying amyloid fibril samples within capillary tubes, drying samples in the presence of a magnetic field to induce alignment, drying samples within a cryo loop used for protein crystallography and drying samples in a hydrated chamber (6–8).

2. Materials

The experiments described here assume users have access to standard laboratory equipment, such as pipettes.

2.1. Dry Stalk Preparation

1. Straight glass capillaries approximately 75 mm in length with an external diameter of 1.4 mm by Proscitech.
2. Glass cutter by Hampton Research.
3. Wax. We have had success using bees wax (available from health food stores) or leg wax (available from a high-street chemist or pharmacists).

4. Glass Petri dish.
5. Hairdryer (see Note 1).
6. Retort stand with clamp.
7. Blu-Tack® by Bostik or other soft removable adhesive.
8. 10 μL of a sample of amyloid fibrils with a concentration of approximately 10 mM (see Note 2).

2.2. Hydrated Sample Preparation

1. A sample of amyloid fibrils for further concentration. The sample should contain at least 2×10^{-6} moles of protein in fibrillar form.
2. EITHER (a) polycarbonate tubes 7 × 20 mm for ultracentrifugation by Beckman and an ultracentrifuge capable of 100,000 × *g* OR (b) centrifugal concentrators: Millipore Amicon Ultra® (alternatively, Microcon®, Centricon®, or Ultrafree® are also appropriate) 100,000 molecular weight cut-off, from Millipore; also a benchtop centrifuge that can accommodate these concentrators.
3. Thin-walled glass capillary tubes with a diameter of 0.7 or 1 mm and wall thickness of 0.01 mm. These are available from various companies, including Hilgenberg in Germany, Hampton Research in the USA, or Capillary Tube Supplies Ltd in the UK.
4. A 50 ml plastic tube.
5. Cotton wool.
6. Adhesive labels.
7. A Bunsen burner or other flame source, such as a match or lighter.

2.3. X-Ray Diffraction Experiments

X-ray scattering patterns are obtained at an X-ray diffraction facility. Many university departments have laboratory-based equipment for X-ray diffraction; scattering experiments can also be performed on X-ray beamlines at national synchrotron facilities. In either case, the setup varies depending on the requirements of the individual beamline. Facilities developed for protein crystallography or fibre diffraction are ideal. In principle, any X-ray diffraction facility can be used, although the quality of the scattering pattern varies depending on the intensity of the X-ray source and the size of the focused X-ray beam. The following is a list of basic requirements for the X-ray facility:

1. A 2D ("area") detector for recording the X-ray scattering pattern (see Note 3).
2. Capability for positioning small samples within the beam (stalks or capillaries need to protrude approximately 0.25 mm into the path of the X-ray beam). An apparatus with a

goniometer for positioning samples and magnifying video camera to visualise the position of samples within the X-ray beam is ideal.

3. Capability to resolve features between 3 and 20 Å (this typically requires the detector to be placed at a distance of 40–300 mm from the sample; see Note 4).
4. Capability to remove or turn off the cryogenic stream that is typically installed on many protein crystallography beamlines. Cooling is not generally required and can generate ice on the stalk.

In addition, when performing X-ray experiments, you will need:

5. A calibrant, such as parafilm or a small piece of high-density polyethylene (HDPE) <1 mm thick, and ~2–5 cm long and 1–3 mm wide.
6. A small quantity of plasticine or Blu-Tack® by Bostik for securing your sample, control and calibrant.
7. A magnetic crystal cap by Hampton Research for loading your sample onto the goniometer. A crystal cap usually used to support cryoloops for protein crystallography is ideal.

You will also need to find out the specific training and safety requirements from the manager of the X-ray facility.

2.4. X-Ray Diffraction Pattern Analysis

1. ImageJ software. This can be downloaded for free from the NIH Web site, at http://rsb.info.nih.gov/ij.
2. X-ray analysis macro for use within ImageJ. This is a text file called YAX_Radial_1_0.txt that can be downloaded from http://www.personal.reading.ac.uk/~scs05ams/YAX/ on Dr A. Squires' research Web page at the University of Reading.
3. Microsoft Excel or equivalent software for manipulating and plotting data.

3. Methods

3.1. Preparing Waxed Capillary Tubes for Dry Stalk Preparation

1. Cut the full length capillary tubes into approximately 50 pieces each roughly 2 cm in length using the glass cutter.
2. Place the wax in the glass Petri dish on the base of the retort stand. Clamp the hairdryer onto the retort stand above the wax. The hairdryer should be at a sufficient distance to the wax so that it gently heats the wax causing it to melt without splashing.

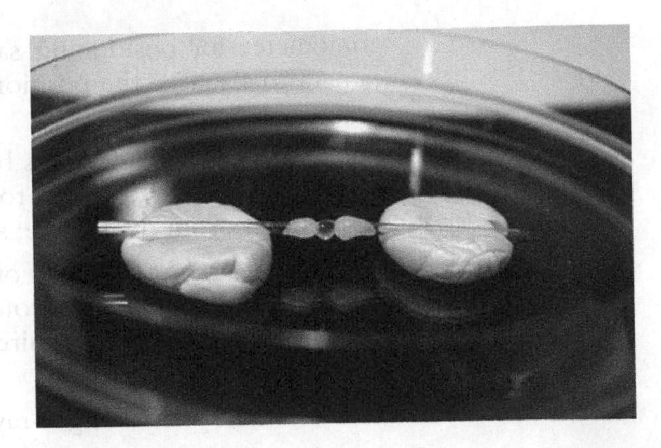

Fig. 1. Experimental equipment for preparing a dried stalk of amyloid fibrils. Two capillaries with wax ends are placed in a Petri dish supported by Blu-Tack® with a droplet of amyloid fibrils suspended between the wax ends.

3. Dip one end of a capillary tube into the molten wax and remove, so there is a small drop of wax at the end of the tube. Remove the capillary and hold upright with the wax at the bottom so that the wax forms a smooth round convex end as it hardens (see the wax ends in Fig. 1; also see Note 5).
4. Repeat the dipping process with the remainder of the capillary tubes. These can be stuck vertically onto the surface of a shelf or a fume hood sash using Blu-Tack® so that they can dry concurrently to form rounded wax ends.

3.2. Dry Stalk Preparation

1. Place two pieces of Blu-Tack® to the base of a second Petri dish and fix two waxed capillary tubes in the Blu-Tack® so that they face each other, end-to-end, with a separation of 3–4 mm between the two wax ends (see separated capillary tubes in Fig. 1). The capillary tubes should be aligned in the same plane if viewed from above or from the side (see Note 6).
2. Using a pipette, place a 10 μL drop of amyloid fibril sample between the two waxed capillary ends, taking care so that the drop does not run up the side of the wax ends and is only in contact with the very end of the wax (see droplet in Fig. 1).
3. Add the lid to the Petri dish and allow the sample to dry between the wax ends. This process typically takes 4–5 h (see Note 7).
4. Once the sample has dried, hold it up to the light to assess the length of the stalk, the thickness and appearance. Stalks should be at least 1 mm in length so that they can be placed in the beam without the wax contributing to the X-ray signal. The thickness is not critical but samples that are clear and look semi-crystalline, as shown in Fig. 2, are more likely to diffract well compared to samples that appear white and lumpy.

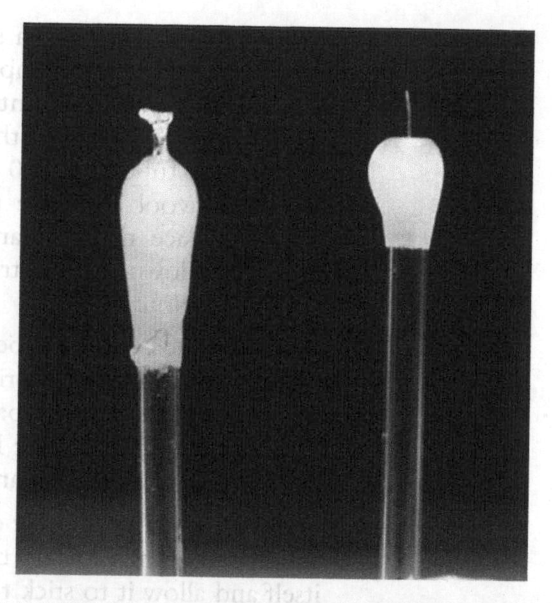

Fig. 2. Two dried stalks illustrating the typical variation in stalk shape for different amyloid fibril samples. Both stalks are relatively transparent and are likely to diffract the X-ray beam well. The X-ray beam should be centred below the flared top of the stalk on the left, as this part of the stalk is less uniform.

Stalks that flare out at the very end are acceptable; try to locate the straight stalk section in the X-ray beam (see Note 8).

3.3. Hydrated Sample Preparation

1. Prepare a concentrated amyloid fibril sample. We have had success concentrating amyloid fibril samples with two different methods:
   (a) Filtering the sample using a Millipore Amicon Ultra® or similar centrifugal filtration unit, following the manufacturer's instructions. This concentration can be carried out with a bench top centrifuge and the fibrils collected from the top of the filter unit for use without further dilution.
   (b) Ultracentrifuging the sample at 100,000 × *g* for 1 h in clear polycarbonate ultracentrifugation tubes. This causes the fibrils to sediment to form a concentrated pellet. The supernatant fluid can be decanted or removed with a pipette and the pellet can be removed and used without further dilution.

   Filtration has the disadvantage that some sample is inevitably lost in the filter and so is not advisable as a method for concentrating very small samples. Ultracentrifugation avoids this problem but centrifuges capable of exerting the forces required are less readily available.
2. Transfer 5–10 μL of the concentrated sample into the wide top of the thin-walled glass capillary. This can usually be done

with a pipette, although a spatula may be required for very viscous or stiff gelled samples. Gentle centrifugation causes the sample to pass down into the thin section of the capillary tube. In order to carry out the centrifugation without breaking the capillary tube, use a 50 ml plastic tube with a small piece of cotton wool or tissue paper placed as padding in the bottom. Place the capillary tube on top of the padding, sealed-end down, and centrifuge at approximately 500 × *g* for 1 min (see Note 9).

3. Next, seal the capillary tube leaving a 2 cm air gap above the sample, by holding it horizontally above a gas flame, and gently pulling the ends apart with a slight twisting motion. The capillary needs to be less than full length to fit within the constraints of the apparatus during X-ray analysis.
4. Carefully fold a small piece of adhesive label around the newly sealed end of the capillary tube. Press the adhesive label onto itself and allow it to stick to the glass but do not press onto the glass, as this will crush the tube. This label allows the sample to be numbered and identified; it is also useful for handling the sample and attaching it using plasticine or Blu-tack® to the crystal cap which is mounted on the goniometer for X-ray analysis (see Subheading 3.4).
5. Prepare a capillary tube containing water (or appropriate solvent blank) in the same manner, to run as a "background" X-ray scattering pattern.

3.4. X-Ray Scattering Experiment

Once at the X-ray source you will need to obtain X-ray scattering patterns from the following samples:

For dried stalks:

1. HDPE or parafilm calibrant.
2. Dried sample stalk.

For hydrated samples:

1. HDPE or parafilm calibrant.
2. Hydrated sample in a sealed capillary.
3. Water or other solvent background in a sealed capillary.

The scattering pattern of the HDPE or parafilm calibrant is required for later software analysis (see Subheading 3.5). This pattern gives the position of the beam centre, the sample to detector distance and the pixel size. If these values are already known accurately, it is possible to carry out the analysis without a calibrant scattering pattern, although the former approach is considerably simpler. You must use the same sample-detector distance for the samples and calibrant. It is better, but not imperative, to use the same exposure time for the hydrated sample and its solvent background. A much shorter exposure time can be used for the calibrant.

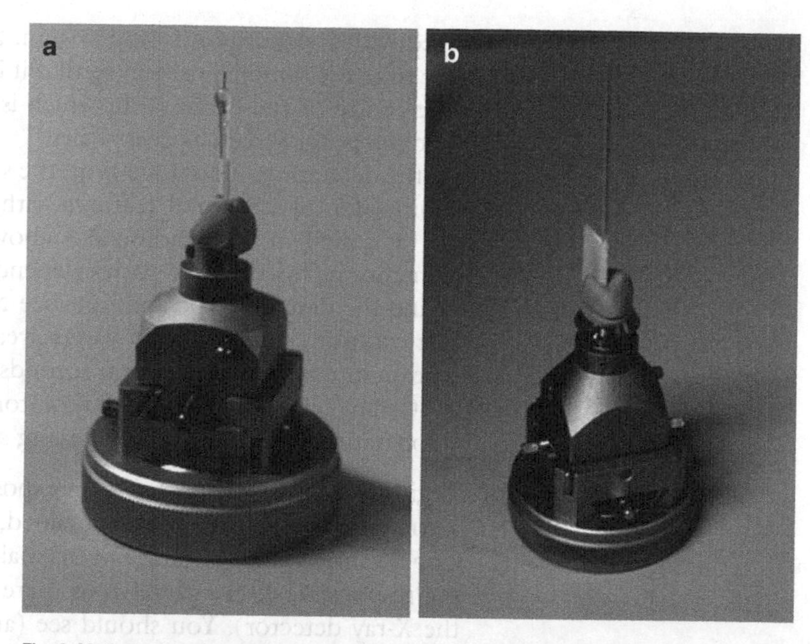

Fig. 3. Methods for mounting amyloid fibril samples on a goniometer prior to X-ray analysis. (a) A dried stalk of fibrils formed on the wax filled end of a capillary tube is mounted on a magnetic crystal cap using Blu-Tack® and then onto a goniometer. (b) Hydrated amyloid fibrils in capillaries can also be mounted onto a crystal cap using Blu-Tack® as shown or threaded through a hole drilled in an old crystal cap before loading on a goniometer. The adhesive label is for handling the fragile capillary and for sample identification.

Before preparing your samples, ask the scientist responsible for the X-ray equipment (beamline scientist) to turn off the cryostream if it is running on the relevant X-ray beamline. This can take 30 min and the equipment may also need to warm to room temperature.

Ask the beamline scientist for advice on how to mount the sample. We have had success mounting dried stalks formed on the end of a wax-filled capillary directly onto an old magnetic crystal cap using either plasticine or Blu-Tack® and then onto a goniometer, as shown in Fig. 3a. A hydrated sample in a sealed capillary can be mounted in a similar way, see Fig. 3b, or if the crystal cap is prepared with a hole drilled through the centre, the capillary can be placed within the hole. Check that the length of your capillary supporting the stalk or containing the hydrated sample is not so long as to cause problems with the range of movement with the goniometer head. The HDPE or parafilm calibrant can be mounted in a similar way and like the samples should be located as close as possible to the centre of this head. Also see Note 10.

Once the sample is mounted, it requires alignment either manually or automatically so that it lies within the path of X-ray beam. For a hydrated sample, the beam should pass through the centre rather than the edge of the capillary tube. If the apparatus is equipped with a video monitoring system, the user should rotate and fine tune the sample position so that it is in focus and aligned with the centre of the beam path, which is often marked on the video or computer screen as cross hairs.

Set the detector to a distance from the sample so that data can be acquired for all structural features with a d-spacing range of 3–20 Å (as stated in Subheading 2.3 above for X-ray beamline requirements). This distance varies depending on the source of X-rays and the detector type and size (see Note 11).

The exposure time required varies greatly depending on the X-ray beam intensity, ranging from seconds (at a synchrotron) up to 10–30 min (on some models of laboratory based copper anode X-ray apparatus). We advise the following sequence:

1. Begin with the calibrant, using an exposure time of 5 s (1 s at a synchrotron). In the image obtained, check that the beam-stop is centred and blocking out the main beam (if not, contact the responsible X-ray scientist, as there is a risk of damage to the X-ray detector). You should see (at least) two concentric rings clearly from the calibrant diffraction pattern. Increase the exposure time until the rings stand out strongly from any background noise. See Fig. 4a, for an X-ray scattering pattern for HDPE. Both HDPE and parafilm give two bright reflections at 4.17 Å (inner reflection) and 3.78 Å (outer reflection).
2. For a dried stalk, again begin with a lower exposure time (e.g. 30 s) to ensure that the beam is hitting the sample. The oriented 4.7 Å inter-strand reflection should be visible in the scattering pattern. Then, increase the exposure time to obtain a better quality image clearly showing inter-strand and inter-sheet reflections that stand out clearly from any background noise. This typically requires ten times the exposure time required for the calibrant. A typical diffraction pattern for a dried stalk of fibrils is given in Fig. 4b.

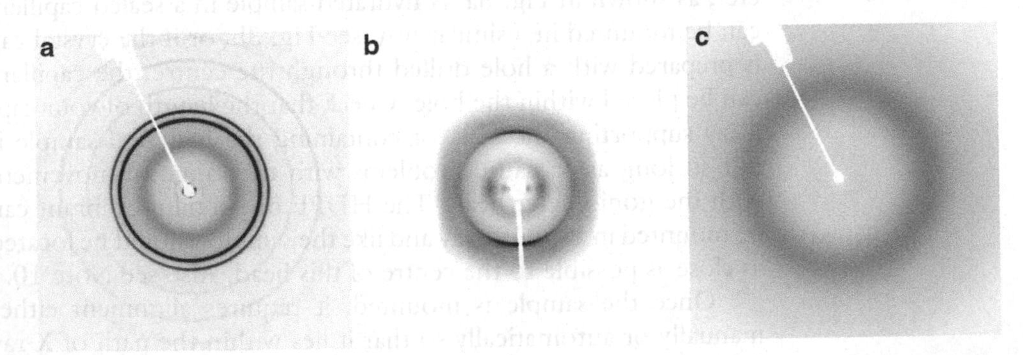

Fig. 4. Typical X-ray scattering patterns for a calibrant, sample and control sample. (**a**) Scattering pattern of HPDE calibrant. (**b**) Scattering pattern for a dried stalk of lysozyme amyloid fibrils in which the fibrils are aligned. (**c**) Scattering pattern for water which is used as a background subtraction for a hydrated sample.

3. For the hydrated sample and solvent background in sealed capillaries, again begin with a lower exposure time (e.g. 30 s) to ensure that the beam is hitting the sample. Look for a broad water reflection at approximately 3.5 Å, as shown in Fig. 4c. Then, increase the exposure time as for dried stalk samples to approximately ten times the exposure time required for the calibrant. The final exposure time should generate a scattering pattern, where the reflection from water appears smooth and stands out from any background noise. Depending on the sample, additional faint rings may also be observed in the pattern collected from the hydrated sample (see ref. 5).

Once you have collected an X-ray scattering pattern for your sample, it is useful to note the following for each pattern: the intensity, the approximate position of reflections (this can usually be determined using in-house software) and whether reflections are aligned. Also take note of the following general parameters: the sample to detector distance, the pixel size, dimensions of the scattering pattern in pixels, the wavelength of the X-ray source (see Note 12), whether your sample is horizontal or vertical in the X-ray beam and the orientation of this axis relative to the final display image (this is dependant on the set-up at different experimental individual facilities).

If possible, export your X-ray scattering data as a 16-bit tiff file (see Note 13) as well as saving the original file format for later analysis of the X-ray scattering patterns as described in Subheading 3.5.

3.5. Analysis of X-Ray Scattering Patterns

Software is required to carry out processing of the 2D X-ray scattering patterns to generate radial intensity profile, to accurately determine the position of reflections and to allow a quantitative comparison between the scattering patterns for dried and hydrated amyloid fibril samples. The radial profile generated by the software presented here is the average intensity of pixels on the scattering pattern as a function of radial distance from the beam centre. We describe how to perform such an integration using macros that we have written, that run within the free software package ImageJ. There is also a range of commercial and free software that can perform more sophisticated analysis (see Note 14), although we have found our macro to be easier to use for basic integration than other packages we tested.

For non-oriented samples, the software gives a single radial profile, averaging over 360° for each radial distance. For aligned samples, the software can be used to generate four radial profiles from wedges extending up, down, left and right from the beam centre that can be orientated in the axial and equatorial direction.

The procedure for obtaining radial profiles from the 2D scattering patterns is as follows:

1. Install ImageJ following instructions from the National Institute of Health Web site (see Subheading 2.4).

2. Next, download the text file for the image analysis macro ("YAX_Radial_1_0.txt") from The University of Reading Web site (see Subheading 2.4) and save it onto your hard drive.
3. Open ImageJ. Go to Plugins>Macros>Install... (or Ctrl+Shift+M), and select the macro file (YAX_Radial_1_0.txt) from the location, where it was installed on your hard drive. It should say "3 macros installed". To check, under Plugins>Macros there should be three new macros called "Calibrate_Manual", "Calibrate_Smart", and "Integrate_Radial Profile".
4. 2D X-ray scattering patterns may be opened and viewed within ImageJ in two ways;
   (a) Some formats can be opened directly from ImageJ using the File>Import>Raw command to import the image file. The exact settings vary depending on the model of X-ray apparatus and sometimes on the computer used. Try importing a pattern with sharp reflections (e.g. the HDPE calibrant). When you are successful, the import settings work with all images obtained on that X-ray apparatus, and ImageJ remembers the settings for future use. For the first attempt, use Image type: 16-bit unsigned, values for image Width and Height (in pixels) noted at the beamline and default settings for all the other import settings; this has proved successful for many detector formats. If the image appears completely black or white, the importing might have been successful but the brightness/contrast may need adjusting. See Note 15 for further approaches if this is not successful.
   (b) Alternatively, convert the scattering pattern into a .tif file, preferably 16-bit rather than 8-bit (see Subheading 3.4 and Note 13). This may be done using the software available on the computer associated with the X-ray apparatus at the facility, where the experiments were carried out. Alternatively, other free software, such as Fit2D, can open a range of X-ray image files and export the data as 16-bit tiff. Within ImageJ, use the File>Open (Ctrl+O) command and select the tif file.
5. Before radial profiles can be obtained, the software needs to have accurate values for the x and y coordinates of the beam centre and the ratio of (sample-detector distance/pixel size). Load a suitable calibrant file (preferably parafilm or HDPE), obtained on the same beamline and with the same sample-detector distance as the sample scattering patterns. Next, identify a bright reflection of known d-spacing (the inner reflection of parafilm or HDPE best). Use the "Points selection" tool and shift-click to select three points in order as follows: (a) the approximate beam centre position (to within ten pixels),

(b) approximately 20 pixels inside the selected reflection and (c) approximately 20 pixels outside the selected reflection. Then, perform the command Plugins>Macros>Calibrate_Smart. The software takes a few seconds to accurately calculate the ring and centre position. It then requires two inputs: the value of the d-spacing for the selected bright reflection and the wavelength of the X-ray beamline. The default values of 4.167 Å and 1.54 Å are the d-spacing of the bright inner ring for HDPE or Parafilm and the wavelength of X-rays emitted by a copper anode source, respectively. If the software has not successfully identified the position of the ring, you can manually position the ring over the reflection and go to Plugins>Macros>Calibrate_Manual. If you do not have a suitable calibrant file, see Note 16.

6. Load your image file, as in step 3. Use the command Plugins>Macros>Integrate_Radial Profile. A dialog box opens. This has the default settings for centreX, centreY and sample-detector distance/pixel size obtained from the Calibration step. For an aligned fibre pattern (e.g. for a dry stalk), leave the "4 wedges" box checked to generate four radial profiles. These profiles correspond to the data from four 30° wedges separated by 90°. By default, these wedges extend up, down, left and right from the beam centre position but these can be tilted using a value for "Wedge Offset". For example, the wedge offset value would be required if the sample stalk was not mounted exactly vertically or horizontally at the time of X-ray diffraction, resulting in the axial and equatorial directions being slightly offset. Leaving the "Total" box checked generates a fifth radial profile that is the average radial intensity measured at each radial point around 360°. This is useful for all scattering patterns. On pressing "OK", the software then carries out the relevant calculations and then opens a dialog box to save a ".txt" file containing the radial profiles. The data is tab delimited (and so can be opened using Excel) with the heading of each column labelled. It contains the d-spacing values in the first column and intensity values for the different radial profiles in subsequent columns. It is useful to plot the radial profile (on *y*-axis) against 1/d (on *x*-axis) because 1/d relates more closely to the scattering angle as seen on the 2D X-ray scattering pattern. For this reason, values for 1/d are output in the second column of the text file.
7. For each subsequent diffraction pattern, repeat step 6. There is no need to recalibrate.

For an aligned pattern, a comparison of the "up", "down", "left" and "right" radial profiles show whether a given reflection is equatorial or axial. Although in principle, only two orthogonal profiles are required, we advise comparing all four profiles for two reasons.

The first is that artefacts can occur resulting in a decrease in intensity on one part of the image, e.g. the beamstop shown in Fig. 4b. The second is that a comparison of peak position confirms the earlier determination of beam centre and gives an indication of error in peak position.

Once you have four radial profiles, if the intensity of the reflections at 4.7 Å reflection in both the "up" and "down" profiles is greater than both the "left" and "right" ones, this is a clear indication of axial alignment consistent with a "cross-beta" structure. This difference in intensity could be expected for a vertical stalk if the amyloid fibrils were aligned. For publication, it is simpler to show only one axial and one equatorial profile or the average of the two axials and two equatorial profiles. An example is shown in Fig. 5a.

A background-subtracted profile from a hydrated sample may be obtained and compared with that from a dried stalk, by following a series of steps outlined below. These steps can easily be performed using a software package, such as Excel.

1. Open the ("total") radial profiles for water and for the hydrated sample.
2. Scale one profile relative to the other so that the 3.5 Å reflections for water lie more or less on top of each other. Even for diffraction patterns taken under identical conditions, we have found that some manual rescaling is needed due to variations in the glass capillary tubes, differences in sample absorbance, etc. An example of water and a hydrated sample scaled ready for subtraction is shown in Fig. 5b.
3. The background-subtracted image for the hydrated sample can then be compared with the total radial profile for the dried stalk, to test for changes in structure on dehydration. For an example, see reference (5).

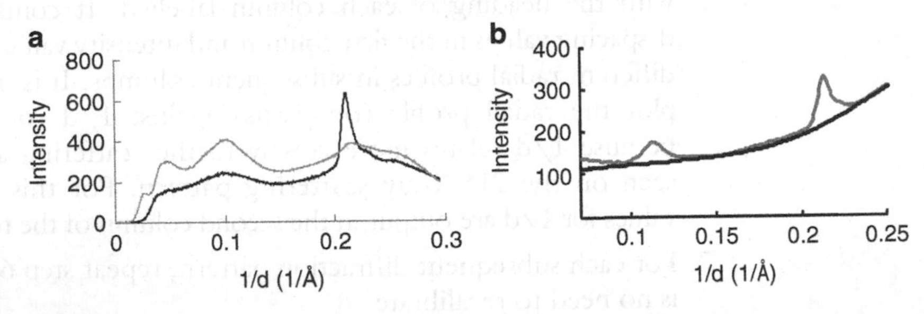

Fig. 5. Radial intensity profiles for dried and hydrated amyloid fibrils. (**a**) The axial (*black series*) and equatorial (*grey series*) radial intensity profiles for dried lysozyme fibrils generated from the aligned scattering pattern in Fig. 4b using the four radial profile function described in Subheading 3.5. (**b**) The average radial intensity profile for hydrated $TTR_{105-115}$ fibrils (*grey series*) generated using the total radial profile function described in Subheading 3.5. The water background (*black series*) is shown rescaled ready for subtraction from the hydrated sample profile.

If your sample is highly aligned, you may also be able to obtain further structural information using other available software packages for the analysis of X-ray scattering patterns (see Note 14).

4. Notes

1. You can use a soldering iron rather than a hairdryer to melt the wax.
2. This sample concentration is approximate; lower concentrations can work with some large proteins, such as insulin and higher concentrations, may be required for some peptides or if only a small fraction of protein is converted amyloid fibrils. It is also likely that you will need to repeat the stalk making process more than once with each sample to obtain a successful stalk. The sample can be viscous but should not be a stiff gel as a gelled sample is unlikely to align.
3. The X-ray detector should ideally be free of any artefacts caused by beam damage.
4. The ability to resolve features to 40 Å can also be beneficial but is dependant on whether the beamstop can be moved further away from the X-ray source.
5. You should avoid long strings of wax which can be removed from the wax end by placing the capillary briefly under the hairdryer; capillaries with concave wax ends should also be discarded.
6. Note that an alternate set-up for the capillaries is a stretch frame, where the distance between the two capillaries can be increased manually as a function of time (8). This can increase the alignment but can also be more labour intensive.
7. If the liquid drop falls from between the capillaries, the set-up should be discarded and new capillary tubes and sample assembled, if the sample looks particularly thin while it is drying between the ends a small amount (~5 μL) of additional sample can be added to the drying stalk.
8. Some amyloid fibril samples may dry so that they remain attached to both wax ends. In this case, gently pull the capillary tubes apart until the amyloid fibril stalk detaches from one of the waxed ends.
9. Highly viscous samples may need to be centrifuged at higher speeds. You may also need to add and centrifuge several sample aliquots depending on how readily the sample flows in the capillary tube.

10. If the beamline lacks a goniometer for mounting the sample, a strip of metal with a hole at one end 2–3 mm in diameter can be inserted vertically in the path of the beam and translated so that the beam passes through the centre of the hole. The dried stalk or capillary can then be mounted using plasticine or Blu-Tack® so that it aligns with the centre of the hole in the metal strip.
11. At a synchrotron X-ray facility, different X-ray beam energies and wavelengths may be available. We have obtained reasonable results using a range of wavelengths from 0.9 to 1.54 Å.
12. The wavelength of X-rays generated using a copper anode source, as found in most laboratory facilities is 1.54 Å.
13. The X-ray scattering pattern data can also be exported as an 8 bit tiff. This file format is more easily opened by a range of imaging software programmes, but the resolution of the data is lower than for the 16-bit tiff. We do not recommend exporting as .jpeg or .jpg, as the compression algorithm involved changes the raw data and can introduce artefacts.
14. Other methods are also available for the analysis of X-ray data, these include: FiberFix (http://www.small-angle.ac.uk/small-angle/Software/FibreFix/FibreFix.html), Clearer (10) and Fit-2D (http://www.esrf.eu/computing/scientific/FIT2D/).
15. If the image resembles "static" on an old-fashioned television, the import has not been successful. In this case, try the following:
    (a) Switch on the "Little-endian byte order" by ticking this box in the import settings.
    (b) If this does not help, you will need to check if you have the right number of bytes per pixel and possibly change "Image Type". First, find the file size in bytes. In Windows, right click on the file name for the scattering pattern and display the file properties. Use the value displayed for "Size", not "Size on disk". The bytes per pixel is calculated by dividing the byte size by the total number of pixels (width × height). If this value is approximately (or slightly over) 2, then try one of the "16-bit" settings, for values around 4 try the different "32-bit" options, and for values around 1 try "8-bit".
    (c) If the image looks reasonable but appears to have been cut in two vertically and rearranged so that the centre of the image is displaced sideways, then you will need to change the value used for "Offset to First Image". To determine this value, calculate the bytes per pixel (use the nearest integer determined in (b) above) × the total number of pixels. Subtract this product from the file size in bytes (use as determined in (b) above). Use this value for the "Offset to First Image" setting.

In each case, remember that if the image appears completely black or white, it is possible that the import was successful and you only need to change the brightness/contrast within ImageJ to observe the reflections.

16. If you do not have a calibrant image, it is still possible to proceed. You will need to have accurate values for the x and y coordinates of the beam centre, and the sample-detector distance in pixels (i.e. the sample-detector distance divided by the pixel size). You will still need to run one of the Calibrate macros so that the software thinks it has values; these values can then be modified when it comes to step 5 of the steps outlined in Subheading 3.5, i.e. the actual manual integration. The procedure is as follows: open any image file, preferably the scattering pattern of a dried stalk or any other image showing sharp reflections. Manually, use the "Elipse" selection tool to position a circle that lies as much as possible on top of a reflection (i.e. matching the exact position of an unaligned reflection, or passing through both arcs for an aligned reflection so that the centre of the new circle approximately matches the beam centre). Use the Plugins>Macros>Calibrate_Manual command. Enter the approximate value for the d-spacing (e.g. 4.7 Å for an amyloid inter-strand reflection). Then, in step 5 (radial integration), change the default values for CentreX, CentreY, and sample-detector distance/pixel size to the appropriate known values.

References

1. Sunde, M., Serpell, L.C., Bartlam, M., Fraser, P.E., Pepys, M.B., Blake, C.C.F. (1997) Common core structure of amyloid fibrils by synchrotron X-ray diffraction. *J Mol Biol.* **273**, 729–39.
2. Wiltzius, J.J.W., Sievers, S.A., Sawaya, M.R., Cascio, D., Popov, D., Riekel, C., et al. (2008) Atomic structure of the cross-beta spine of islet amyloid polypeptide (amylin). *Protein Sci.* **17**, 1467–74.
3. Castelletto, V., Hamley, I.W. (2007) Beta-Lactoglobulin Fibers under Capillary Flow. *Biomacromolecules.* **8**, 77–83.
4. Vestergaard, B., Groenning, M., Roessle, M., Kastrup, J.S., de Weert, M.V., Flink, J.M., et al. (2007) A Helical Structural Nucleus Is the Primary Elongating Unit of Insulin Amyloid Fibrils. *PLoS Biology.* **5**, e134.
5. Squires, A.M., Devlin, G.L., Gras, S.L., Tickler, A.K., MacPhee, C.E., Dobson, C.M. (2006) Ray scattering study of the effect of hydration on the cross-beta structure of amyloid fibrils. *J Am Chem Soc.* **128**, 11738–9.
6. Kendall, A., Stubbs, G. (2006) Oriented sols for fiber diffraction from limited quantities or hazardous materials. *J Appl Crystallogr.* **39**, 39–41.
7. McDonald, M., Kendalla, A., Tanaka, M., Weissman, J.S., Stubbs, G. (2008) Enclosed chambers for humidity control and sample containment in fiber diffraction. *J Appl Crystallogr.* **41**, 206–9.
8. Makin, O.S., Serpell, L.C. (2005) X-Ray diffraction studies of amyloid fibrils, in *Methods in Molecular Biology.* (Sigurdsson, E. M., ed.), Humana, Totowa, NJ, pp. 67–80.
9. Kishimoto, A., Hasegawa, K., Suzuki, H., Taguchi, H., Namba, K., Yoshida, M. (2004) beta-Helix is a likely core structure of yeast prion Sup35 amyloid fibers. *Biochem Biophys Res Commun.* **315**, 739–45.
10. Makin, O.S., Sikorski, P., Serpell, L.C. (2007) CLEARER: a new tool for the analysis of X-ray fibre diffraction patterns and diffraction simulation from atomic structural models. *J Appl Crystallogr.* **40**, 966–72.

Chapter 11

Solid-State NMR of Amyloid Membrane Interactions

John D. Gehman and Frances Separovic

Abstract

Solid-state NMR pulse sequences often feature fewer pulses and delays than the more common solution NMR experiments. This ostensible simplicity, however, belies the care with which experimental parameters must be determined, as solid-state NMR can be much less forgiving of improper experimental set-up. This is especially true of "semi-solid" samples, such as the phospholipid vesicles used to study membrane-associated peptides and proteins, which feature prominently in misfolding diseases. Protocols for the preparation of multilamellar vesicles for solid-state NMR studies of Aβ peptides are described, together with procedures for optimization of critical experimental parameters, such as spectral widths, delay times, and field strengths for ^{31}P, ^{2}H, and ^{13}C NMR spectroscopy.

Key words: Solid-state NMR, Lipid–peptide interactions, Phospholipid membranes, Amyloid peptide, Alzheimer's disease

1. Introduction

Membranes and misfolded proteins are each a bane of traditional structural biology. At a molecular level, both of these targets typically are large, insoluble, and possess insufficient long-range order, which limit the potential of solution nuclear magnetic resonance (NMR) and crystal diffraction. It seems especially perverse, then, that accumulating evidence suggests both may be at the core of two common forms of dementia. Both Parkinson's and Alzheimer's diseases are characterized by misfolding of polypeptide chains, and increasingly their basis for disruption of nerve cell function and cytotoxicity appears linked to mechanical and/or chemical perturbation of cellular membranes.

Solid-state NMR, meanwhile, has emerged as a strong complement to traditional approaches in structural biology, as it is not restricted by the same sample requirements as solution-state

Andrew F. Hill et al. (eds.), *Protein Folding, Misfolding, and Disease: Methods and Protocols*, Methods in Molecular Biology, vol. 752, DOI 10.1007/978-1-60327-223-0_11, © Springer Science+Business Media, LLC 2011

NMR and crystal diffraction. Consequently, peptides may be studied in a more appropriate environment than the detergents and solvents often employed by these traditional strategies to address structural questions of membrane polypeptides. Solid-state NMR also offers an additional advantage, as perturbations to lipid components of the membrane bilayer by the polypeptide can be studied directly.

The use of solid-state NMR to study such "semi-solid" or liquid crystalline samples is not without its own limitations, however. Specific isotopic enrichment of molecules involved in the lipid/polypeptide complex is frequently necessary, and significant sample quantities are required to provide sufficient signal-to-noise in a reasonable experimental time. Samples are consequently expensive, making optimized sample preparation and instrumental set-up imperative to gain the best information without excessive usage of labelled materials, specialized molecular components and NMR time. The optimizations described in this chapter include the lipid/polypeptide ratio, radio frequency (RF) field strengths, pulse sequence delay times, and other solid-state NMR acquisition parameters. Our protocol descriptions pertain to studies of the amyloid peptides from Alzheimer's disease, Aβ, in phospholipid bilayer systems (1–3).

2. Materials

Aβ peptides used in our studies were produced by solid-phase synthesis and studied in mixed phospholipid bilayers of palmitoyloleoylphosphatidylcholine (POPC) and palmitoyloleoylphosphatidylserine (POPS). For deuterium NMR studies, POPC was used with the palmitoyl chain deuterated (d_{31}-POPC), which was obtained from Avanti Polar Lipids (Alabaster, AL).

2.1. Solid-State NMR

1. For samples to be run only by static NMR, inexpensive liquids NMR tubes of an appropriate diameter can be cut down and flame-sealed and polished on either end.
2. Magic angle spinning (MAS) rotor assemblies require careful attention for use with peptide/lipid vesicle suspensions of varying water content. Where sample oxidation is of particular concern, it is important for rotor assemblies to be air-tight. In any case, some rotor assemblies expose the sample to the dried air supplies used for MAS such that moist samples dehydrate during the experiment. It is often difficult to take advantage of the full sample volume with truly air-tight assemblies, and the leakage past the ostensible seals in nominal liquids assembly rotors suggest that they are not air tight. In our experience, MAS rotors should be underfilled by about 25% when using lipid dispersions.

3. The rotor assembly and cap materials should be chosen with care, e.g. Torlon components unfortunately give background for $^{13}C\{^{1}H\}$ CP MAS in the spectral region shared by carbonyl ^{13}C, and often with greater intensity than that from limited sample quantities. Vespel and PMMA are other common materials used for rotor components, but also suffer, from $^{13}C\{^{1}H\}$ background signal. Teflon or Kel-F are preferred for CP experiments, but cannot be used for high-speed MAS.

3. Methods

3.1. Peptide/Lipid Complex

3.1.1. Multilamellar Vesicles with "Associated" Peptide

1. Lyophilized Aβ peptide is dissolved at ~1 mg/mL in hexafluoroisopropanol (HFIP), and warmed at 25°C for 30–60 min to break up aggregates.
2. HFIP is removed by rotary evaporation in a round-bottom flask (usually 25 or 50 mL volume), leaving a thin film of peptide.
3. Lipids (30:1 lipid/peptide mol ratio) are dissolved to ~20 mg/mL in chloroform/methanol (9:1 v/v) and similarly dried by rotary evaporation into a thin film in a separate round-bottom flask (see Note 1).
4. After pumping the flasks at low vacuum for several hours to remove traces of solvent, multilamellar vesicles (MLVs) are hydrated with buffer (see Note 2), transferred to a microcentrifuge tube, and are subjected to four freeze/thaw cycles (liquid N_2/tepid water).
5. MLV suspension is then added to the thin peptide film, and swirled gently over the peptide film (as for the MLV preparation itself) until film is removed.
6. Samples which are to contain a large water excess may be used as is, if sample is not too dilute. Otherwise, samples may be benchtop-centrifuged below the gel–liquid phase transition temperature and some supernatant removed (for constant buffer concentration), and/or lyophilized to a gel-like consistency (whereby buffer concentration increases). A mass of water equal to that of lipid may be added back to the lyophilized lipid/peptide sample to provide the minimum amount of water considered to be fully hydrated (see Note 3).

3.1.2. MLV with "Incorporated" Peptide

1. Lyophilized Aβ peptide is dissolved in HFIP and dried as above (Subheading 3.1.1, steps 1 and 2).
2. Lipids are dissolved to ~20 mg/mL in chloroform/methanol (9:1 v/v) and added to the peptide film (i.e. in the same flask) then rotary evaporated to a thin film (see Note 1). The flask is pumped at low vacuum for several hours to remove trace solvents.

3. MLV and lipid are hydrated together and freeze/thawed as above (Subheading 3.1.1, step 4).
4. Final sample preparation as above (Subheading 3.1.1, step 6).

3.1.3. Addition of Metal Ions

1. 0.1 M metal(II) glycinate stock solutions are prepared by combining $CuCl_2$ or $ZnCl_2$ with two equivalents of glycine.
2. Appropriate volumes of metal ion stock solution are added to an empty microcentrifuge tube; MLV/peptide suspensions are transferred to the droplet of metal ion solution between the first two and last two freeze/thaw cycles (as ~1–2 μL of relatively concentrated stock metal ion is added, and a concentrated pocket of metal in the lipid/peptide sample should be avoided for fear of nucleating aggregation).

3.2. Optimization of Static NMR Experiments

The NMR spectrometer uses low-frequency (audio) filters to reduce noise entering the receiver by reducing the bandwidth of noise to the spectral region of interest, which is usually implied by the specified spectral width. These audio filters are imperfect devices, however, and are known to introduce spectral artefacts. The sudden impact of signal on the audio filters causes ringing, which distorts the first few points of the free induction decay (FID), and causes misshaped spectral baselines. There is also a lag in initial response to the sudden appearance of signal, which manifests as a different phase shift for different frequencies (4, 5). The use of echo pulse sequences to collect static spectra permits the simultaneous optimization of pre-acquisition delays which obviate the difficult adjustment of linear phase corrections for wide-line spectra, and permit the audio filters to be opened early, prior to when the echo is formed, so that the distorted first few points of the FID may be avoided. Properly optimized delays in echo sequences provide for flat baselines and true lineshapes, which otherwise can be difficult to discern in wideline spectra.

3.2.1. ^{31}P Pre-acquisition Delay (see Note 4)

1. Use a Hahn echo pulse sequence with $90^{\circ}_{\varphi1} - \tau_1 - 180^{\circ}_{\varphi2} - \tau_2 - Acq_{\varphi3}$ with $\varphi_1 = xy\overline{xy}xy\overline{xyxy}xy$, $\varphi_2 = yxyx\overline{yxyx}xyxy\overline{xyxy}$ (6) and $\varphi_3 = xy\overline{xy}$ (see Note 4). Using 85% H_3PO_4 to measure the field strengths (see Note 5) with adequate recycle delay (2 s at 121.5 MHz ^{31}P frequency), determine power necessary for a minimum field strength equivalent to 250 ppm for ^{31}P to avoid off-resonance complications. Typically, a 5 μs 90° pulse is adequate.
2. Set full spectral width (sw) such that the dwell time (dw = 1/sw in quadrature detection) can be achieved by the signal digitizer with no rounding error, and provides for ~10 oversampling of the rapidly decaying FID. For example, a ~50 ppm wide static ^{31}P line runs from approximately 33 to −17 ppm

with the carrier set to the isotropic chemical shift. A minimum spectral width of ~8 kHz is required on a 300 MHz instrument. If a minimum dwell time increment of only 1 μs resolution is assumed, 62.5 or 100 kHz spectral widths are used. These provide ~10 times the number of points for the short lifetime of the FID and correspond to 16.0 and 10.0 μs dwell times, respectively.

3. Note the on-resonance frequency for the 85% H_3PO_4 ^{31}P chemical shift reference sample. This frequency is the reference chemical shift of 0 ppm.
4. Set the delay between the 90° and 180° pulses to 5 × dw, and the total of any delays prior to opening the receiver to 1 × dw. Set the delay between opening of the audio receiver and the start of digitization to $1/(gd \times v_c)$, where gd is the group delay factor for a particular audio filter and v_c is the audio filter cut-off frequency. The gd factor may need to be empirically determined, and is further optimized here. For example, using a spectral width of 62.5 kHz, 5 μs 90 time, and elliptical audio filters set to a cut-off frequency of 34 kHz, the pulse sequence is initially set to (2s) – (5 μs 90° pulse) – (80 μs delay) – (10 μs 180° pulse) – (16 μs delay) – (receiver on) – ((1.29 × 34 kHz)$^{-1}$ delay) – (Acq).
5. Acquire a series of 1D spectra in which the carrier offset is arrayed from on-resonance to +50 ppm or more, in 2,000 Hz increments (or less), as in Fig. 1. Use a sufficient number of scans to give suitable signal to noise ratio, and at a multiple of the phase cycle (i.e. $n \times 16$).
6. Left-shift the resulting FIDs by four points prior to Fourier transformation. These left shifts compensate for beginning acquisition earlier than the echo maximum, and at the same time discards data which is contaminated by the transient response of the audio filters. This provides for much more flat spectral baselines underneath the broad static ^{31}P lineshapes. Set the frequency-dependent linear phase correction (φ_ω) to zero, and adjust the frequency-independent zero-order phase correction (φ_0) for the on-resonance spectrum to give an absorptive peak. Call this phase adjustment φ_a, and note the frequency difference $\Delta\omega_a$ between the on-resonance peak and the phase pivot frequency (often the right or left edge of the spectrum). Apply this zero-order phase adjustment to all other spectra in the array. Identify a second spectrum in which the peak is <180° out of phase with the on-resonance peak. Keep the frequency-dependent linear phase correction at zero, and adjust the frequency-independent zero-order phase correction for the second peak; call this phase adjustment φ_b,

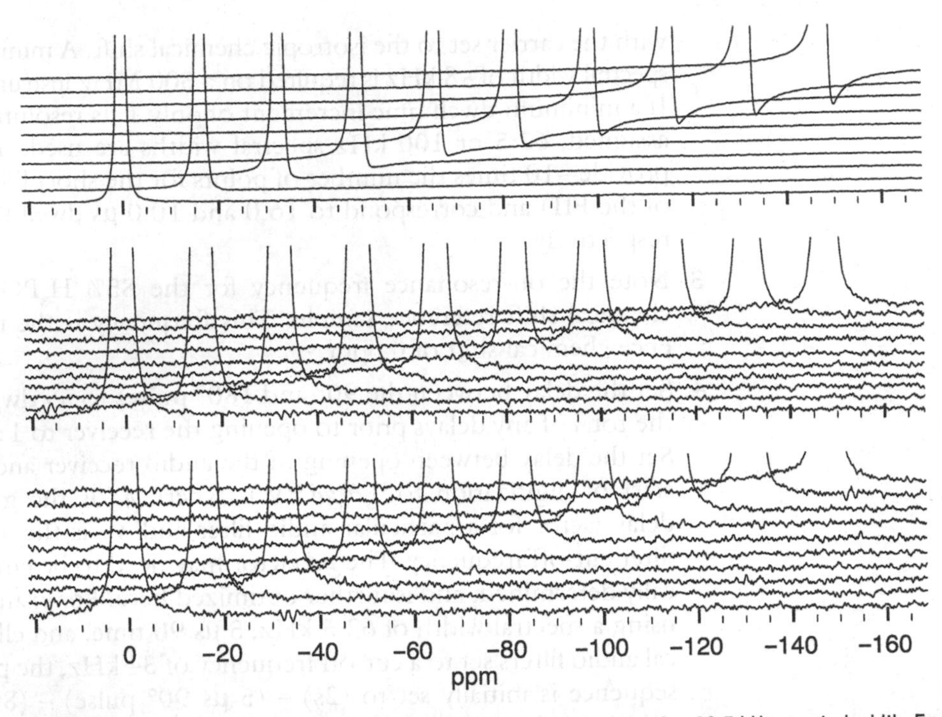

Fig. 1. Optimization of pre-acquisition delay for the [31]P Hahn echo experiment using 62.5 kHz spectral width. Each panel shows a spectral series in which the carrier frequency if offset such that an 85% H_3PO_4 sample peak moves from on-resonance to far off resonance. (*Top*) The spectral series resulting from an initial guess at the correct pre-acquisition delay with zero frequency-dependent phase correction. Inspection of zero-order phase corrections required to phase different spectra in the series indicates a 52.25° frequency-dependent phase correction is necessary if the phase pivot point is the right edge of the spectra. (*Middle*) Setting this phase together with zero-order phase readjustment gives correct baselines for the same data. (*Bottom*) A corresponding 2.33 μs adjustment to the pre-acquisition delay together with zero frequency-dependent phase adjustment provides for properly phased spectra with flat baselines across a spectral region which far exceeds that which the ^{31}P line will appear. Magnification is increased on the bottom two panels to stress the phasing at baseline. Note, however, that the baselines for the full width spectra in the middle panel are curved as a consequence of the frequency-dependent phase correction.

and note the frequency $\Delta\omega_b$ analogous to $\Delta\omega_a$. Calculate φ_0 and φ_ω using

$$\varphi_\omega = \frac{\mathrm{sw}\cdot(\varphi_b - \varphi_a)}{\Delta\omega_b - \Delta\omega_a}$$

$$\varphi_0 = \varphi_a - \frac{\omega_a}{\mathrm{sw}} \bullet \varphi_\omega = \varphi_b - \frac{\omega_b}{\mathrm{sw}} \bullet \varphi_\omega$$

7. Confirm that these phase corrections provide for absorptive peaks across the array of spectra. Adjust the digitization delay

$$\Delta\tau = \frac{10^6}{\mathrm{sw}}\frac{\varphi_\omega}{360}$$

and repeat from step 4 until peaks which appear at the frequency range over which the static lipid ^{31}P phospholipid line will appear are absorptive with first-order phase correction set to zero (see Fig. 2).

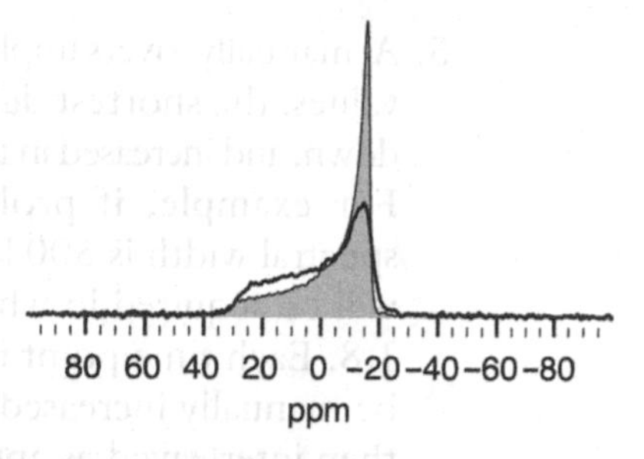

Fig. 2. Static ^{31}P signal from 100% dimyristoylphosphatidylcholine (DMPC) MLV in the fluid lamellar phase (30°C) collected on a 300 MHz Varian Inova spectrometer. The spectrum arises from vesicles in which lipids have aligned in the magnetic field (see Note 7), and the open spectrum is a typical true ^{31}P lineshape with blunted features owing to T_2 relaxation.

3.2.2. ^2H Pre-acquisition Delay (see Note 6)

1. Each lipid chain of the diacyl phospholipid experiences different dynamics; interpretation of results during optimization is simplified by using MLV of single chain-deuterated lipids.
2. The solid/quadrupolar echo pulse sequence ($90°_{xy\bar{x}\bar{y}} - \tau_1 - 90°_{y\bar{x}\bar{y}x} - \tau_2 - \mathrm{Acq}_{xy\bar{x}\bar{y}}$) should be configured using a ^2H field strength ω_{2H} greater than half the expected quadrupolar splitting Q, and τ_1 delay less than double the inverse of the quadrupolar splitting (7, 8).

$$\omega_{2H} > \frac{Q}{2} \quad \text{and} \quad \tau_1 \leq \frac{2}{Q}$$

The echo is expected at $2\tau_1 - (8\omega_{2H})^{-1}$, assuming two 90° pulses. Alternatively, composite pulses specific for three-level systems may be employed for improved excitation bandwidth (9).
3. The most prominent feature of ^2H spectra of perdeuterated lipid acyl chains is either an isotropic HDO resonance or the 90° edge of the CD_3 deuteron Pake pattern (see Note 7). The integral of the spectral intensity is dominated by the more numerous and more ordered CD_2 deuterons, which differ slightly in chemical shift from the HDO and CD_3. The center of the spectrum must be set to the center of the CD_2 quadrupolar splittings, when signal-to-noise becomes sufficient to reveal resolved features for different CD_2 positions.
4. As for the static ^{31}P experiment, frequency-dependent spectral phase corrections cannot be relied upon to provide correct lineshapes. Set the spectral width to the maximum permitted by the digitizer. Set audio filter cut-off frequency to approximately 1.1 times the maximum expected quadrupolar splitting to reduce the noise and avoid audio filter jitter which occurs at maximum bandwidth settings.

5. A manually oversampled data set is obtained by arraying τ_2 values; the shortest delay is the minimum to avoid pulse ringdown, and increased in ten steps with increment $0.1\times dw=0.1/sw$. For example, if probe ringdown requires 5 μs, and the spectral width is 500 kHz (dwell time is 2 μs), ten 1D spectra will be acquired in which $\tau_2 = 5 + t$ where t=0, 0.2, 0.4, ..., 1.8. Each time point in the corresponding FIDs should then be manually increased by t. Data from each of these FIDs is then interleaved as appropriate to form a single FID with (in this example) 5 MHz spectral width.
6. With the echo in the FID more clearly defined, an appropriate number of points is left-shifted before downsampling the data back to an appropriate spectral width. The optimum number of left-shifts should be identified from examination of the solid echo formed in the FID and trial of processing (Fourier Transform and line broadening) configurations, which should be inspected for baseline artefacts as in Fig. 3.
7. The optimum τ_2 delay to be used henceforth is determined by the optimum number of left-shifts in step 5. For example, if pulse ringdown required 5 μs delay, and 330 left-shifts provides the most artefact-free spectrum from the 10× manually oversampled 5 MHz data set, future spectra should be collected with τ=6.8 μs and 33 left-shifts using 500 kHz spectral width.
8. Static ^{2}H spectra of deuterium-enriched phospholipids are often dePaked (10, 11) to transform the unoriented powder spectrum into a 0°-oriented spectrum when comparison of order parameter details throughout the lipid acyl chain are desired.

3.3. Optimization of (CP)MAS Experiments

3.3.1. Optimization of ^{31}P MAS Experiments

While static ^{31}P linewidths may address headgroup orientation and/or perturbation to lipid headgroup order on the 10^{-4} s timescale, spin-lattice (T_1), and transverse (T_2) relaxation measurements address perturbation to motion on the 10^{-9} and 10^{-3} s timescales, respectively. While relaxation times vary for different lipid orientations within the vesicle, average values may be measured much more quickly, and different lipid components frequently differentiated, by narrowing peak intensities to the isotropic positions using MAS. The experiments themselves are relatively straightforward, e.g. the inversion recovery and Hahn echo sequences for T_1 and T_2, respectively. There is some limitation to the MAS rate which may be used (see Note 8). Care must also be taken in the case of the Hahn echo to ensure that each half of the echo delay is an integer multiple of the spinning speed. This necessitates programming of correct pulse sequence times and good control of stable MAS spin rate.

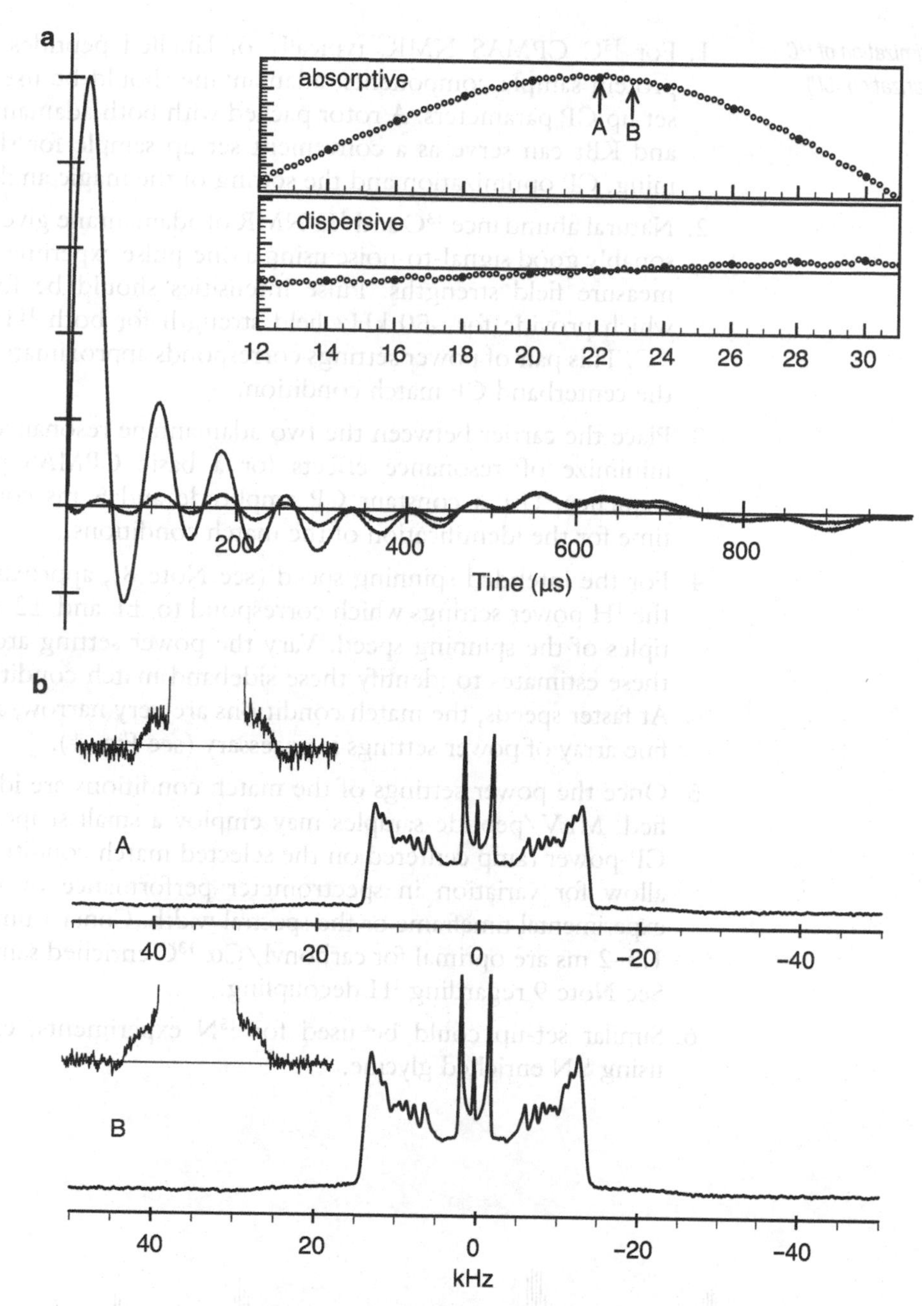

Fig. 3. Static ²H data for d_{54}-DMPC MLV in the fluid lamellar phase (30°C) collected on a 300 MHz Varian Inova spectrometer. (**a**) The FID with 0.2 μs dwell time constructed from interleaving ten 2 μs FIDs, between which the pre-acquisition delay was incremented by 0.2 μs. (**b**) Fourier transformed spectra with magnified baselines inset, using FIDs left shifted to points A and B as indicated in (**a**). Point A represents the top of the echo, and should be the ideal place to begin Fourier transformation. Insufficient ²H field strength (61 kHz), notwithstanding composite pulses, causes some baseline distortion in spectral regions far from the center of the spectrum. Left-shifting to point B minimizes baseline distortions, at the expense of missing the echo maximum. See Davis (7) for thorough discussion.

3.3.2. Optimization of ^{13}C Cross-Polarization (CP)

1. For ^{13}C CPMAS NMR, typically of labelled peptides and protein sample components, adamantane should be used to set up CP parameters. A rotor packed with both adamantane and KBr can serve as a convenient set-up sample for shimming, CP optimization and the setting of the magic angle.
2. Natural abundance ^{13}C and ^{1}H NMR of adamantane give reasonably good signal-to-noise using a one-pulse experiment to measure field strengths. Pulse intensities should be found which provide for ~50 kHz field strength for both ^{1}H and ^{13}C. This pair of power settings corresponds approximately to the centerband CP match condition.
3. Place the carrier between the two adamantane resonances to minimize off-resonance effects for a basic CPMAS pulse sequence. Use a constant CP amplitude and 5 ms contact time for the identification of the match conditions.
4. For the intended spinning speed (see Note 8), approximate the ^{1}H power settings which correspond to ±1 and ±2 multiples of the spinning speed. Vary the power setting around these estimates to identify these sideband match conditions. At faster speeds, the match conditions are very narrow, and a fine array of power settings is necessary (see Fig. 4).
5. Once the power settings of the match conditions are identified, MLV/peptide samples may employ a small shaped ^{1}H CP-power ramp centered on the selected match condition to allow for variation in spectrometer performance over the experimental timeframe or the spectral width. Contact times of 1.5–2 ms are optimal for carbonyl/Cα-^{13}C enriched samples. See Note 9 regarding ^{1}H decoupling.
6. Similar set-up could be used for ^{15}N experiments, except using ^{15}N enriched glycine.

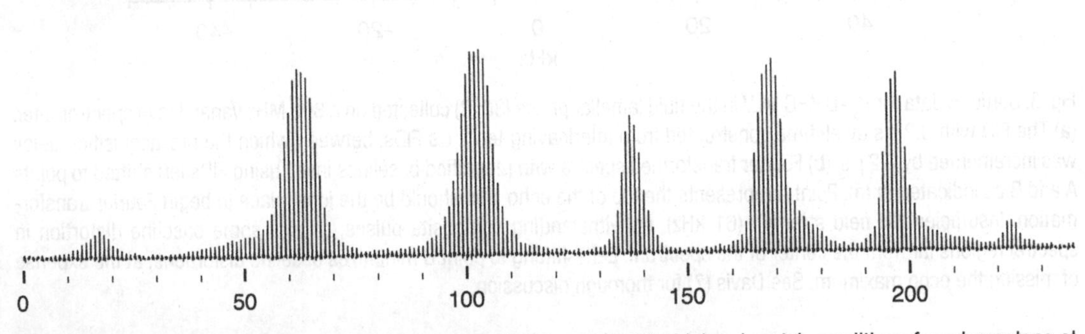

Fig. 4. Cross-polarization match profiles from the −3 (*left*) to +3 (*right*) sideband match conditions for adamantane at 5 kHz MAS rate on a 600 MHz Varian VNMRS spectrometer using Varian BioMAS 3.2 mm scroll coil T3 probe with ^{1}H power arrayed in 238 nominal steps of 0.023 dB. Only the CH_2 peak is shown.

4. Notes

1. Additional molecular mass of one water molecule should be added to the nominal molecular mass of phospholipids. As even unlabelled peptide can be very expensive, a single sample preparation is typically used for static ^{31}P, ^{2}H, and ^{31}P MAS relaxation experiments. Consequently, ^{2}H-enriched acyl chain lipid is often incorporated at ⅓ – ½ of the total lipid.
2. It is best to hydrate lipids and form MLV first with half of the buffer volume, taking care that the buffer solution washes over the lipid film only. The lipid film should come up into the thickening solution only slowly, using manual swirling, intermittently warming with the heat of the hand, and a rotary-mixer at the lowest setting. When the first half of the buffer volume appears to have picked up most of the lipid, transfer it to a microcentrifuge tube, and use the second half of the buffer volume to collect the remainder of the lipid from the flask. Each lipid mixture behaves differently; if additional rinses are necessary to clean up the remainder of the lipid, small successive volumes of water may be used. Lipid suspensions are difficult to transfer with plastic pipette tips and thin glass Pasteur pipettes. We use glass capillaries with Teflon plunger tips from Drummond Scientific (Broomall, PA) available through local distributors.
3. The final stages of sample treatment are governed largely by the volumes available for the solid-state NMR experiment. Samples to be used only for ^{31}P and ^{2}H can generally be more dilute, as typical appropriate probe coils can accommodate 5 mm OD and larger, cut and polished glass NMR tubes. MAS rotor sample volumes are typically 3.2–7 mm, with useful payload volumes for semi-solid samples ranging from 22 to 130 μL.
4. The present generation spectrometers no longer mix radio frequencies down to audio frequencies before digitizing the signal, and consequently no longer use the hardware audio filters which require consideration of group delay. Pre-acquisition delays are, therefore, determined mostly by pulse ring-down times. A large number of spectrometers in operation, however, do require audio filter-dependent optimization of pre-acquisition delays. These audio filters are also restricted with respect to maximum bandwidth, often 256 kHz. Note also that the phase cycle provided is particularly forgiving of improper 90 and 180° tip angles; changes in spectrometer performance over time is not obvious using this 16-step phase cycle. This provides for true static ^{31}P lineshapes, but belies a potentially narrowing excitation profile.

Field strength calibration with H_3PO_4 should, therefore, be conducted semi-regularly.

5. Field strengths should be calibrated by pulsing on-resonance, and identifying the characteristic 360° pulse length time. This requires more signal-to-noise than a 90° pulse to identify clearly, but is more forgiving with respect to underestimated recycle delay times.
6. For purposes of illustration, we discuss optimization of the static ^{2}H static quadrupolar echo experiment for a sample with maximum width of the superimposed CD_n Pake doublets of typically in the range 60–100 kHz.
7. Lipids tend to align with their acyl chains perpendicular to the magnetic field, creating static lineshapes that gives more intensity at the 90° edge and decreasing intensity though to the 0° edge. This effect can be minimized for samples to be studied in the liquid lamellar phase by equilibrating the sample above the gel–fluid phase transition temperature outside the magnetic field.
8. MAS experiments should be optimized using spinning rates with a rotor period determined by the time resolution of the NMR console, in anticipation of pulse sequences which require rotor-synchronized pulses. For example, 2.5, 4, 5, and 6.25 kHz MAS rates correspond to 400, 250, 200, and 160 μs rotor periods which would prevent the accumulation of rounding errors in rotor-synchronized pulse sequence timing. Many spectrometers are capable of better time-resolution than whole μs, in which case additional MAS rates also allow for exactly timed rotor periods. MAS rates for vesicles are limited; in our experience, MLV prepared in excess water can be unstable (depending on composition) in 5 mm rotors at speeds greater than 5 kHz while samples with ~30 water molecules per lipid spun in 3.2 mm rotors have been stable up to 8 kHz.
9. Many ^{1}H decoupling schemes exist. In our experience, CW decoupling is sufficient at 300 MHz for fluid phase lipids, provided the decoupling field strength is not equal to small integer multiples of the spinning speed (12). At higher field strengths, SPINAL64 (13) provides better decoupling performance, with ±10 ppm bandwidth at 600 MHz. SPINAL64 is a 64-step pulse phase cycle, which is optimized by adjusting the tip angle of each pulse around 165° and the governing phase value around 15°. Decoupler center frequency and bandwidth can be simultaneously determined by setting the Cα carbon of glycine (enriched if possible when using smaller rotor volumes) on-resonance and varying the decoupler frequency offset.

Acknowledgments

The authors would like to thank Dr Tong Lay Lau for helpful discussions regarding the sample preparation procedure and the Australian Research Council for financial support.

References

1. Lau, T.-L., Gehman, J. D., Wade, J. D., Masters, C. L., and Barnham, K. J. (2007) Cholesterol and Clioquinol modulation of Aβ(1–42) interaction with phospholipid bilayers and metals. *Biochim. Biophys. Acta* **1768**, 3135–3144.
2. Lau, T.-L., Gehman, J. D., Wade, J. D., Perez, K., Masters, C. L., and Barnham, K. J. (2007). Membrane interactions and the effect of metal ions of the amyloidogenic fragment Aβ(25–35) in comparison to Aβ(1–42). *Biochim. Biophys. Acta* **1768**, 2400–2408.
3. Gehman, J. D., O'Brien, C. C., Shabanpoor, F., Wade, J. D., and Separovic, F. (2008) Metal effects on the membrane interactions of amyloid-beta peptides. *Eur. Biophys. J.* **37**, 333–44.
4. Schaefer, J., and Stejskal, E.O. (1974) Baseline artifacts in high-resolution Fourier transform NMR spectra. *J. Magn. Reson.* **15**, 173–176.
5. Hoult, D. I., Chen, C.-N., Eden, H., and Eden, M. (1983) Elimination of baseline artifacts in spectra and their integrals. *J. Magn. Reson.* **51**, 110–117.
6. Rance, M., and Byrd, R. A. (1983) Obtaining high-fidelity spin-1/2 powder spectra in anisotropic media: phase-cycled Hahn echo spectroscopy. *J. Magn. Reson.* **52**, 221–240.
7. Davis, J. H. (1983) The description of membrane lipid conformation, order and dynamics by 2H-NMR. *Biochim. Biophys. Acta* **737**, 117–171.
8. Schmidt-Rohr, K., and Spiess H. W. (1994) *Multidimensional solid-state NMR and polymers.* Academic Press, San Diego, CA.
9. Levitt, M. H., Suter, D., and Ernst, R. R. (1984) Composite pulse excitation in three-level systems. *J. Chem. Phys.* **80**, 3064–3068.
10. Bloom, M., Davis, J. H., and MacKay, A.L. (1981) Direct determination of the oriented sample NMR spectrum from the powder spectrum for systems with local axis symmetry. *Chem. Phys. Lett.* **80**, 198–202.
11. Sternin, E., Bloom, M., and MacKay, A.L. (1983) De-Pake-ing of NMR spectra. *J. Magn Reson.* **55**, 274–282.
12. Mehta A. K., Tounge, B. A., Burns, S. T., Gehman, J. D., Wu I., Coker, G. S., Pomerantz A. E., Paulson, E., Luptak, A., and Zilm, K. W. (1997) CPMAS with High Speed Spinning: Resolution, Decoupling and Recoupling. *39th Rocky Mount. Conf. Anal. Chem.*
13. Fung, B. M., Khitrin, A. K., and Ermolaev K. (2000) An improved broadband decoupling sequence for liquid crystals and solids. *J. Magn. Reson.* **142**, 97–101.

Chapter 12

Sedimentation Velocity Analysis of Amyloid Fibrils

Chi Le Lan Pham, Yee-Foong Mok, and Geoffrey J. Howlett

Abstract

Analytical ultracentrifugation is a classical technique used to study the solution behavior of proteins. Experimentally determined sedimentation coefficients provide information regarding the size, shape, and interactions of biological macromolecules. Sedimentation velocity methods have been used to characterize the different aggregation states of amyloid oligomers and fibrils. This chapter first describes the theoretical background for sedimentation velocity analysis. It then provides experimental protocols for sedimentation velocity experiments using the analytical ultracentrifuge. Finally, this chapter describes the procedure used to analyze sedimentation velocity data to obtain the size distribution of amyloid fibrils and their oligomeric precursors.

Key words: Analytical ultracentrifugation, Sedimentation velocity, Size distribution, Sedimentation coefficient, Amyloid fibrils, Oligomers

1. Introduction

Amyloid fibril formation is a multistep process that involves the initial self-association of protein into small soluble oligomers and protofibrils that elongate into mature fibrils. Traditionally, the pathology and progression of amyloid-related diseases is attributed to the accumulation of large amounts of amyloid fibrils into insoluble deposits that disrupt the normal functioning of tissues and organs. This deposition process in vivo is accompanied by the accumulation of nonfibrillar components, such as serum amyloid P (SAP), apolipoprotein E (apoE), proteoglycans, and lipid (1). Nonfibrillar components influence the interactions and tangling of amyloid fibrils (2) and have the potential to exert regulatory effects on both innate immune and proteolytic surveillance mechanisms (3). While large amyloid deposits may be pathogenic for systemic

Andrew F. Hill et al. (eds.), *Protein Folding, Misfolding, and Disease: Methods and Protocols*, Methods in Molecular Biology, vol. 752, DOI 10.1007/978-1-60327-223-0_12,

amyloidosis (4), studies suggest that small soluble oligomers or protofibrils on the pathway to amyloid fibril formation are more pathogenic than the insoluble amyloid fibrils (5–10). These early intermediates of fibril formation are cytotoxic in cell culture and are implicated in a number of amyloid-related diseases, particularly in neurodegenerative diseases, such as Alzheimer's disease and Parkinson's disease (9–11).

To thoroughly understand the mechanism of amyloidosis, better information is required on the structure and interactions of the various intermediates and mature fibrils. However, separation and structural characterization of amyloid fibrils and their solution behavior has proved to be challenging because of the extreme size and insoluble nature of amyloid fibrils. In recent years, analytical ultracentrifugation techniques have to be used to study the solution properties of amyloid fibrils and their oligomeric intermediates.

1.1. Analytical Ultracentrifugation

Analytical ultracentrifugation involves monitoring the concentration distribution of a solute during sedimentation and is a well-established technique used for studying the self-association and interactions of protein in solution. Analysis of sedimentation velocity behavior offers a number of advantages over other classic procedures for analyzing interacting systems, such as dynamic light scattering, size exclusion chromatography, or electrophoresis. For particles of constant friction coefficient, sedimentation rates depend on the (2/3) power of the molar mass. This is in contrast to diffusion-based methods, such as chromatography and dynamic light scattering, which depend on the Stokes radius and where separation is based on the (1/3) power of the mass. This explains the vastly increased resolution of sedimentation velocity analysis compared with these other techniques. Furthermore, sedimentation velocity analysis has a firm theoretical basis; involves no matrices, surfaces, or bulk flow; and is extremely versatile with respect to the size ranges of the interacting species under consideration.

The two main techniques in analytical ultracentrifugation are sedimentation equilibrium and sedimentation velocity. Both techniques use identical instrumentation but differing experimental protocols. In sedimentation equilibrium experiments, short solution columns and low speeds are used to ensure that equilibrium is reached. Analysis of the time invariant concentration distribution yields information about molecular masses and interactions. A major advantage of this approach is that the analysis of equilibrium gradients makes no assumptions regarding hydrodynamics or molecular shape. Sedimentation equilibrium methods have been useful in defining the molecular nature of the subunits that form amyloid fibrils (12–14). However, the large sizes of amyloid fibrils, in general, prevent their analysis by sedimentation equilibrium because the fibrils fail to form a measurable equilibrium concentration gradient and quickly sediment to the bottom of the cell.

A significant advantage of sedimentation velocity, on the other hand, is the capacity to analyze large fibrillar structures. Sedimentation velocity monitors the radial concentration distributions (boundaries) of solutes during transport through a centrifugation field over a period of time. Figure 1a is a schematic diagram of a typical sedimentation velocity profile showing the position of the reference and the sample meniscus. The sedimentation boundary is the region between the solvent plateau (the region where solute is depleting from the sample meniscus) and the solution plateau (the region of maximum concentration

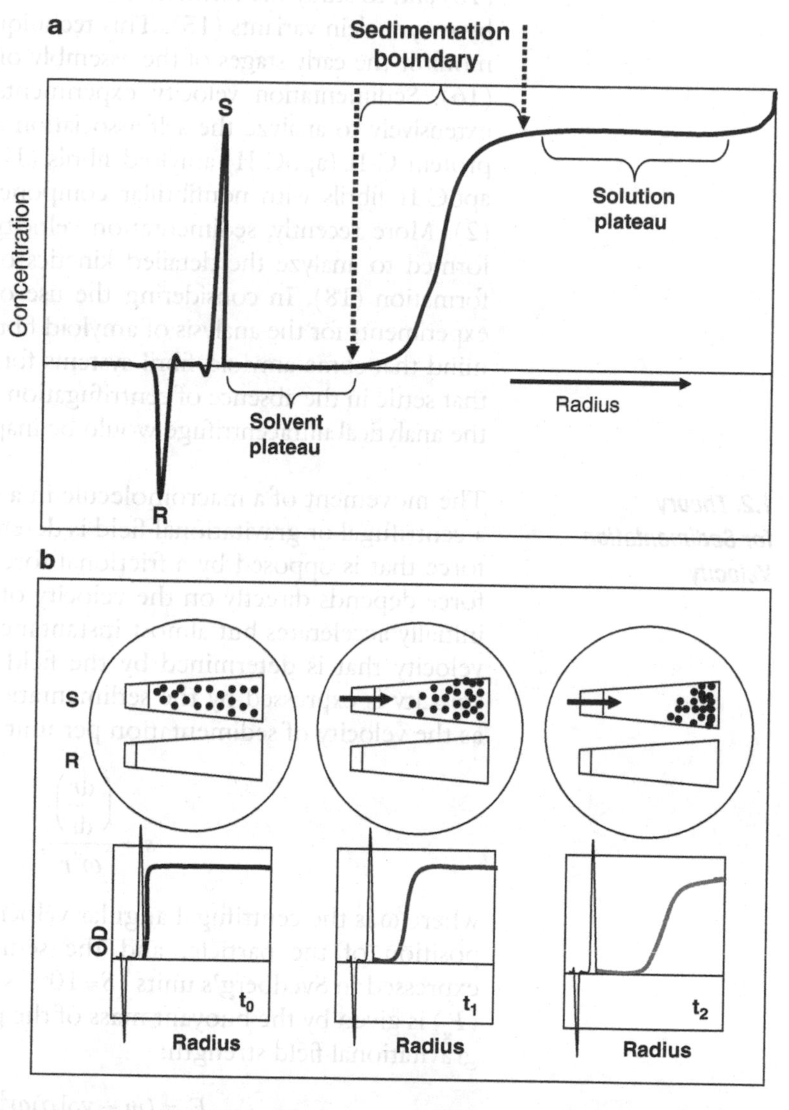

Fig. 1. Illustration of a typical sedimentation velocity experiment. Panel (**a**): A schematic diagram of a typical sedimentation velocity profile showing the position of the reference meniscus (R) and the sample meniscus (S). The sedimentation boundary is the region between the solvent and the solution plateau. Panel (**b**): An illustration of a sedimentation velocity experiment showing the different sedimentation profiles observed at different stages of sedimentation.

distribution). Figure 1b is an illustration of sedimentation velocity experiments showing a typical double-sector cell containing the protein sample and the appropriate reference. At the start of the experiments, (t_0), the protein concentration distribution along the radial position of the cell is uniform. At various times (i.e., t_1, t_2) of sedimentation, at constant speed and temperature, the boundary moves down the cell as the solute is depleted from the sample meniscus.

Sedimentation velocity methods have been used to characterize the quaternary structure of the fibril forming protein transthyretin (13) and to study the formation of small oligomers and protofibrils by α-synuclein variants (15). This technique has also been used to monitor the early stages of the assembly of β_2-microglobulin fibrils (16). Sedimentation velocity experiments have been performed extensively to analyze the self-association and tangling of apolipoprotein C-II (apoC-II) amyloid fibrils (17) and the interaction of apoC-II fibrils with nonfibrillar components of amyloid deposits (2). More recently, sedimentation velocity experiments were performed to analyze the detailed kinetics of apoC-II amyloid fibril formation (18). In considering the use of sedimentation velocity experiments for the analysis of amyloid fibrils, it should be borne in mind that some amyloid fibril systems form flocculent precipitates that settle in the absence of centrifugation. In such cases, the use of the analytical ultracentrifuge would be inappropriate.

1.2. Theory for Sedimentation Velocity

The movement of a macromolecule in a uniform solution within a centrifugal or gravitational field is determined by a gravitational force that is opposed by a frictional force. Because the frictional force depends directly on the velocity of movement, the particle initially accelerates but almost instantaneously reaches a constant velocity that is determined by the field strength. This constant velocity is expressed by the sedimentation coefficient (s) defined as the velocity of sedimentation per unit field strength:

$$s = \frac{\left(\frac{\mathrm{d}r}{\mathrm{d}t}\right)}{\omega^2 r}, \tag{1}$$

where ω is the centrifugal angular velocity (rad/s), r is the radial position of the particle, and the sedimentation coefficient is expressed in Svedberg's units ($S = 10^{-13}$ s). The gravitational force (F_g) is given by the buoyant mass of the particle multiplied by the gravitational field strength:

$$F_g = (m - \mathrm{vol}\rho)\omega^2 r, \tag{2}$$

where the buoyant mass of the particle is defined as the mass of the particle (m) reduced by the mass of the displaced fluid, given as the product of the particle volume (vol) and the solution density (ρ).

The opposing frictional force (F_f) is directly proportional to the velocity of the particle and related by a frictional coefficient f as follows:

$$F_f = -f\frac{dr}{dt}. \quad (3)$$

Equating F_g and F_f and combining Eqs. 1–3 for a macromolecule of molecular mass (M) and partial specific volume ($\bar{v}$) (in mL/g) yields:

$$s = \frac{\left(\frac{d\ln r}{dt}\right)}{\omega^2} = \frac{M(1-\bar{v}\rho)}{Nf}, \quad (4)$$

where N is Avogadro's number. Equation 4 relates the sedimentation coefficient of a sedimenting species to its molecular weight, partial specific volume, and the frictional coefficient. Thus, macromolecules with different molecular weights, partial specific volumes, and frictional coefficients move at different velocities and yield different sedimentation coefficients in a given solution density and centrifugal field. Implicit in Eq. 4 is the dependence of the sedimentation coefficient on solution viscosity η, which affects the frictional coefficient. Experimentally determined values for sedimentation coefficients are generally corrected to standard condition of pure water at 20°C as:

$$s_{20,w} = s\left(\frac{(1-\bar{v}\rho)_{T,b}}{(1-\bar{v}\rho)_{20,w}}\right)\left(\frac{\eta_{20,w}}{\eta_{T,b}}\right), \quad (5)$$

where the subscript 20, w denotes the value for water at 20°C and the subscript T, b denotes the experimental conditions.

Equation 4 can be integrated to calculate the time required (t) for a species of known s to sediment from the sample meniscus, r_m, to the bottom of the cell, r_b, for any chosen angular velocity:

$$\ln r_b - \ln r_m = s\omega^2 t. \quad (6)$$

The evolution of a concentration distribution, $c(r, t)$, for a single species of diffusing particle as a function of time is given by the Lamm equation (19):

$$\frac{dc}{dt} = -\frac{1}{r}\frac{d}{dr}\left[rD\frac{dc}{dr} - s\omega^2 r^2 c\right], \quad (7)$$

where D is the diffusion coefficient and c is the concentration.

Analysis of experimental data in terms of both s and D has until recently been dependent on approximate solutions of Eq. 7 (19–21) or on graphic transformations of the raw data (22). Although these approaches have proved valuable, they have imposed significant limitations on experimental design and on the levels of sample complexity that can be addressed. More recently, new methods have been developed based on fitting numerical simulation of the Lamm equation to the experimental data using iterative nonlinear regression methods (23–25). One application of this approach is the c(s) method which enables direct fitting of the experimental concentration distributions to a model comprising multiple noninteracting species to yield a best fit sedimentation coefficient distribution (23). This approach has been applied in the analysis of the polymerization of α1-antitrypsin to form dimers, trimers, and higher order oligomers (26).

For analyzing the sedimentation behavior of large particles, such as polymers or amyloid fibrils, where diffusion is negligible, simpler analytical procedures pertain. The evolution of the concentration distribution of a population of identical nondiffusing particles with a uniform initial distribution is given by step functions:

$$U(r,t) = e^{-2\omega^2 st} \times \begin{cases} 0 & \text{for } r < r_m e\omega^2 st \\ 1 & \text{else} \end{cases} . \quad (8)$$

Most notable among the methods used to analyze the sedimentation velocity behavior of nondiffusing particles is the recently developed ls-g*(s) approach in which experimental concentration distributions are modeled in a least-squares fashion by a summation of step functions weighted according to a variable distribution of sedimentation coefficients (27). Analogous distributions are also available from time-derivative analysis methods in which apparent sedimentation coefficient distributions are obtained as a transform of the time-derivative of the experimental concentration distributions (28, 29). These approaches are especially valuable for the analysis of amyloid fibrils for which the assumption of no diffusion is a good approximation.

Both the c(s) and ls-g*(s) procedures are encapsulated in a computer program, SEDFIT, for analyzing sedimentation data (23, 25, 27, 30, 31). This program is freely available (http://www.analyticalultracentrifugation.com). The Web site is, in addition, an excellent resource for the information on the different approaches used in the analysis of sedimentation data for both diffusing and nondiffusing particles using the c(s), ls-g*(s), and the VanHolde–Weischet method.

2. Materials

2.1. Instrumentation

1. All analytical ultracentrifugation experiments are performed using the optima™ XL-A or XL-I analytical centrifuge equipped with a 4-hole or an 8-hole rotor.

2. *Rotor*: For each of the rotors, one of the holes is required for the counterbalance, with its reference holes that provide calibration of radial distance. Therefore, sedimentation experiments can be performed for three protein samples (in a 4-hole rotor) or seven samples (8-hole rotor) simultaneously under the same centrifugation conditions (same speed and temperature). This is useful when performing a concentration dependent study on protein self-association.
3. *Double-sector cell*: A double-sector cell consists of a centerpiece with two sectors (Fig. 1b) sandwiched between two circular quartz windows. This assembly is held in place by an aluminum casing. One sector compartment contains the protein sample while the other sector (reference sector) is for sample buffer. This type of assembly allows the passage of light through the sector cavities for the measurement of sample concentration. The use of the double-sector cell allows the measured absorbance to be corrected for background contribution due to sample buffer.

2.2. Protein Samples

Protein samples should be of high purity solubilized in a buffer system of choice. However, the buffering system should be non-absorbing, and should not contain any component that may interfere with the absorbance at the wavelength used in the experiment. For example, DTT absorbs at 280 nm when oxidized while sodium azide absorbs strongly at 230 nm. Therefore, it is essential to avoid the addition of DTT and sodium azide in the sample buffer when performing experiments at 280 or 230 nm, respectively. The protein concentration for sedimentation velocity experiments should have an absorbance in the range of 0.1–1.5.

2.3. Reference

The reference solution should contain the same buffer composition as the protein samples.

3. Methods

3.1. Experimental Procedures

1. Tighten the double-sector cell(s) using a torque wrench to 120–140 psi. Exceeding this limit may warp the casing and cause the quartz windows to break. The tightening of the cell is to ensure that there is no leakage from the sample or reference sectors.
2. Using gel loading pipette tips, load 380 μL of the protein solution into the sample sector and 400 μL of buffer into the reference sector of the cell (see Note 1).
3. Seal the cells with the small filling hole screws.

4. Insert and align the cells into the holes of the rotor with the filling holes facing the center of rotation. If a balance is required, a double-sector cell is filled with buffer solution in both the sample and reference sector.
5. Place the rotor in the centrifuge chamber. Tighten the monochromator (optical arm) into place.
6. Leave the rotor in the centrifuge under vacuum for approximately 30–60 min to equilibrate the temperature of the samples (see Note 2).
7. From the computer, open the software program that controls the data acquisition of the centrifuge. Select the type of rotor you are using (4-hole or 8-hole rotor) and set the temperature to 20°C (see Note 3).
8. In the "Detail" button in the information box for each of the cell, select the type of centerpiece that you are using (i.e., double sector=2). Enter the computer directory you wish your data to be saved in (if information is the same for all the samples in the experiment, simply enter the detail for cell 1 only and check the "all cells identical to cell 1" from the main screen).
9. Perform a radial calibration at 3,000 rpm (650g). Select velocity and absorbance mode. In the "option" menu, check "radial calibration before first scan" and click on "single scan" button in the main screen. Once the calibration is complete, uncheck the radial calibration in the "option" menu.
10. Perform an initial wavelength scan at 3,000 rpm (650g). Select wavelength scan to measure the absorbance between 200 and 300 nm at a radial distance of 6.5 cm (in the middle of the cells). In the "Detail" option, enter the number of measurements you wish to average (i.e., 5) and click on "single scan" button in the main screen. Review the initial wavelength scan of all the cells and decide on a single wavelength scan that would be most appropriate to perform the experiment. Note that the absorbance of each sample needs to be in a range of 0.1–1.5.
11. At the chosen wavelength, perform an initial radial scan at 3,000 rpm (650g). For each sample in the experiment, select velocity and absorbance mode. Scan radial position from the top of the cell, R_{min} = 5.8, to the bottom of the cell, R_{max} = 7.2. In the "Detail" menu, type 0.03 radial step size (for fast scans) in continuous mode with 1 replicate. Decide the radial range you wish to collect each radial scan. Usually, data is collected from 0.1 cm before the reference meniscus to the bottom of cell (R_{max} = 7.2 cm). The initial radical scan allows you to test for any leakage between the sample and reference sector (see Note 4).

12. Set up a method run by:
    (a) Enter the rotor speed. The speed used in the experiment depends on the size of the protein to be studied. For example, sedimentation velocity experiments of monomeric apoC-II (8.9 kDa) are performed at 40,000 rpm (116,000g) while experiments with apoC-II amyloid fibrils are performed with a rotor speed of 4,000–8,000 rpm (1160–4600g). The program SEDFIT allows the predicted behavior of a particular species to be calculated as a guide for choosing an appropriate speed.
    (b) Set the radial distance where you want to start and finish the collection of each radial scan (R_{min} and R_{max}).
    (c) For each cell, select the velocity and absorbance option and set the wavelength for the collection of data. In the "Detail" menu, select 0.002 radical step size (for high resolution) in continuous mode with 1 replicate.
    (d) From the "Method" menu, set the time interval between each radical scan and the number of scans to be collected and enter "0" for "delay start" to start the method with no delay. The aim is to collect scans until the sedimentation boundary has cleared the meniscus and all of the protein molecules in solution (or most) have sedimented to the bottom of the cell. Usually, the instrument takes approximately 90 s to collect a radical scan from one cell. Therefore, if you are performing experiments with a single sample, the minimum time interval between radical scans should be approximately 2 min, and if you have three samples in an experiment, the minimum time interval would be approximately 5–6 min. If a higher rotor speed is used in the experiments, a shorter time interval is required to collect data.
    (e) In the "Option" menu, check "stop XL after last scan" and enter the number of scans you want to overlay during the running of the experiment.
    (f) Start "Method scan" from the main screen.
    (g) The experiment normally takes approximately 2–4 h to perform depending on the size of the protein and the speed used in the experiment.

An example of data obtained from sedimentation velocity experiments is shown in Fig. 2. Here, sedimentation velocity analysis was performed to monitor the kinetics of fibril formation by apoC-II. ApoC-II was refolded in 0.1 M sodium phosphate buffer, pH 7.4 at a concentration of 0.4 mg/mL and fibril formation was monitored by sedimentation velocity analysis after 10, 25, and 240 h of incubation at 20°C (18). The sedimentation velocity experiments were performed at a speed of 8,000 rpm (4600g).

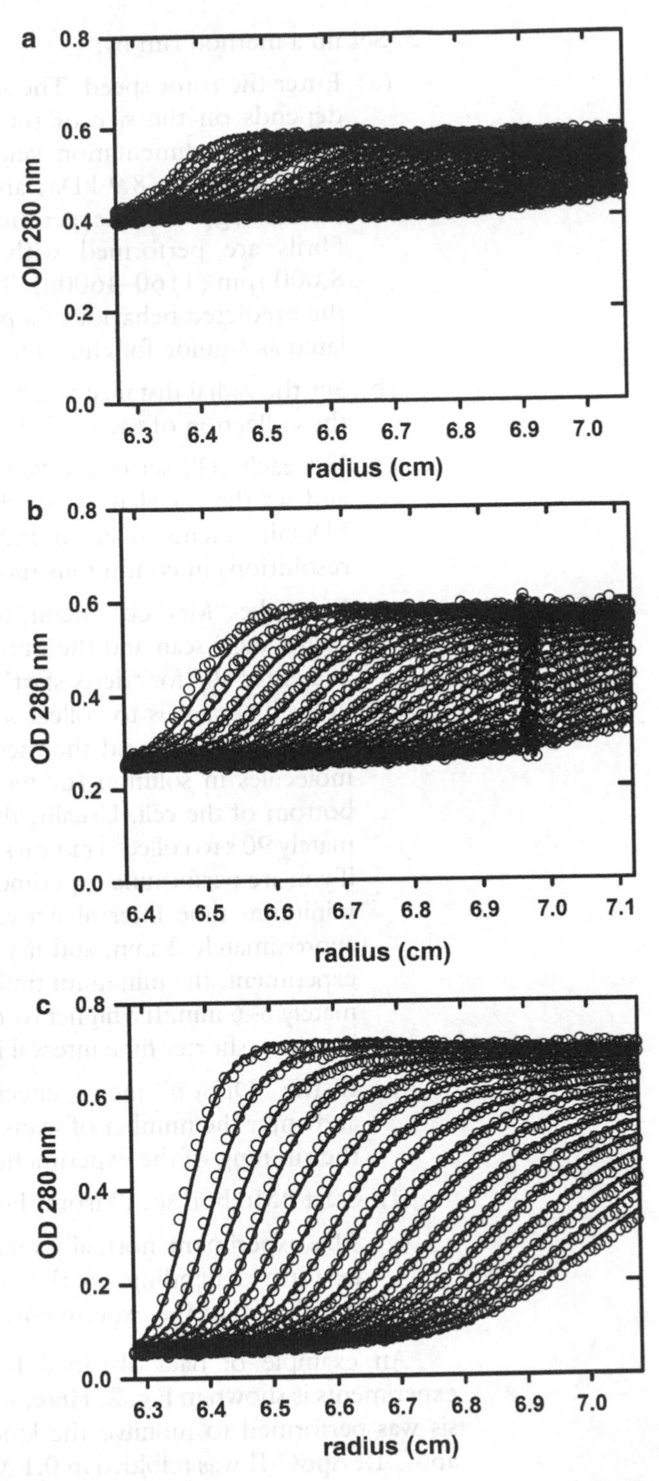

Fig. 2. Time-dependent sedimentation velocity analysis of apoC-II amyloid fibrils formed at 0.4 mg/mL after 10 h (**a**), 25 h (**b**), and 240 h (**c**) incubation at 20°C. Ultracentrifugation was performed at 8,000–10,000 rpm (4600–7280g) with radial scans collected every 8 min for 320 min. The theoretical fits to the experimental data using the c(s) model are shown as *solid lines*. For clarity, only every second scan is shown. Reproduced from Binger et al. (18) with permission from the publisher.

Radial scans were collected at 280 nm, in 8-min intervals for 320 min. As shown in Fig. 2, after 10 h incubation only a small fraction of material is in the fast moving boundary with a high nonsedimenting baseline evident. The nonsedimenting OD is approximately 70% of the starting OD, suggesting that most of the apoC-II is monomeric. After 25 h, a larger proportion of apoC-II is in the fast moving boundary with only 40% is nonsedimenting. After 240 h incubation, only 8% of the initial OD is nonsedimenting suggesting that 92% of the protein has formed fibrils.

It should be noted that sedimentation velocity experiments can be performed using interference optics instead of absorbance optics. This inference option is available on the XL-I model. In this case, it is important to match the meniscus closely to avoid artifacts by the sedimentation of buffer. Interference optics offer the advantage that scans can be collected quickly (less than 10 s) with high accuracy and sensitivity and are useful for solution conditions in which there is interference from strongly absorbing ligands or buffer components.

3.2. Data Analysis

Sedimentation velocity profiles can be analyzed using the computer program called SEDFIT. This program allows sedimentation boundaries to be analyzed by various models, including the c(s) (23) and ls-g*(s) (27) models, described above in the theory Subheading 1.2. The Web site: http://www.analyticalultracentrifugation.com has a step-by-step tutorial for size distribution analysis. It is highly recommended that the reader goes through the step-by-step tutorial prior to analyzing their data for the first time.

The fitting of sedimentation velocity profiles involves choosing appropriate fitting parameters. In practice, start the fitting process by choosing a wide range and then narrowing the selection. The ls-g*(s) model is often used in the analysis of the size distribution of amyloid fibrils, where diffusion is negligible. However, a recent study (18) has used the c(s) model to analyze the size distribution of apoC-II amyloid fibril to closely examine the presence of short fibrils and closed-loop structures. The following is the general procedure for fitting sedimentation velocity data to the c(s) and ls-g*(s) model:

1. Begin fitting the data by loading approximately 40 scans covering the complete sedimentation process. If more than 40 scans are collected, you may choose to load every "nth" scan.
2. Upon loading the data, the radial position of the sample meniscus and the bottom of the cell need to be specified. The position of the sample meniscus cannot be determined with absolute precision from the experimental data using either absorbance or interference data and can be included in the fitting procedure to obtain optimal fits to the data (see Note 5). Limits on the range of meniscus position to be used in the fitting procedure can be applied.

3. Select an appropriate model for the analysis of sedimentation velocity data (e.g., c(s) or ls-g*(s)).
4. The regularization methods allow fitting data using a predefined confidence level. There are two methods available in SEDFIT for regularization. The first regularization method is known as the "second derivative Tikhonov–Phillips" approach (27). This method of regularization is used for samples that are expected to have a broad distribution, i.e., polymers. The second method is known as "maximum entropy" and is often used when fitting data for a mixture of discrete macromolecules (23) (see Note 6). You can choose the regularization method under the "option" drop-down menu and select the "size distribution option."
5. Set fitting parameters: (1) sedimentation coefficient increments (i.e., resolution); increasing the number of increments gives a better fit to the experimental data but can lead to significant oscillations in the distribution; (2) smallest and largest sedimentation coefficient values (*s*-min and *s*-max, respectively); setting and then adjusting the *s*-min and *s*-max values can improve the fitting results; (3) confidence level (*F*-ratio) also known as the regularization parameter (*P*-parameter); the setting of the *P*-parameter determines the amount of regularization used and also defines the statistical confidence level of the fitted distribution.
6. For c(s) analysis, enter the frictional ratio (f/f_0) of your protein, i.e., $f/f_0 = 1.0$ for globular proteins. The value of f/f_0 can be fitted along with the sample meniscus. Buffer density and viscosity along with partial specific volume need to be entered. These parameters can be estimated using the computer software SEDNTERP (freely available for download from http://www.jphilo.mailway.com).
7. Algorithms are included in SEDFIT to minimize time-independent and radial-independent noise. Time-independent noise may arise from optical imperfections in the cell windows, whereas radial-independent noise arises as a result of the entire scan moving up or down at different time intervals. This latter effect is particularly relevant to data collected using interference optics and is attributed to "fringe jitter" (see Note 7).
8. Rather than minimizing time-independent and radial-independent noise, it is also possible to fit the data assuming a single value for the baseline nonsedimenting absorbance or to fix the baseline correction to an experimentally determined value obtained at higher angular velocities.
9. An option is also available in SEDFIT to fit the data using logarithmic (log spaced s grid) rather than linear spacing. This approach gives better coverage of sedimentation coefficients in the lower range.

10. Once all the parameters are entered, select "Run" from the menu. The computer program fits the data to the parameter set. The goodness of the fit to the experimental data is shown as solid lines on the upper graph while the residuals are shown in the middle of the graph and the fitted sedimentation coefficient distribution is shown on the bottom graph. The root mean squared deviation (rmsd) value is displayed on the screen.
11. Inspect the fitted results firstly by looking at the residuals map and the rmsd value. Repeat the fitting procedure by adjusting the fitting parameters (i.e., increased the resolution and/or adjusting the *s*-min and *s*-max) to obtain the best of fit to your experimental data.
12. To fit the position of the meniscus and f/f_0 value, in the parameter box, check the appropriate box and select "Fit" from the menu.
13. For a good fit to the experimental data, (1) the fitted data (solid line, upper graph) should follow that of the experimental sedimentation velocity profiles, (2) the residual map should be random and evenly distributed, and (3) the rmsd should be consistent with the random noise in the data.
14. SEDFIT has an option for calculating the weight-average sedimentation coefficient from the fitted distribution.
15. You can copy the raw data, the fitted data, the residual data, and the distribution data into a clipboard and paste it into a worksheet in Microsoft Excel, Sigmaplot, or other graphical programs.

In addition to the c(s) and ls-g*(s) model, SEDFIT has a number of other model options that may be used to analyze the sedimentation velocity behavior of oligomeric species and amyloid fibrils:

(a) *The VanHolde and Weischet approach* (22). The approach has been used to characterize the distribution of amyloidogenic states of transthyretin, Aβ, and α-synuclein (13–15). In this approach, each scan of the sedimenting boundary is divided into equal concentration fractions and the position of each fraction is converted to apparent sedimentation coefficients. These are then calculated for sequential scans and combined with an extrapolation to infinite time to give the sedimentation coefficients of the fractions corrected for the effects of diffusion. For samples of mixed species, only boundary divisions that originate from positions in the solution in which the sample is homogeneous yield true apparent *s*-values (31), and as such, the sedimentation coefficient distribution should only be qualitatively interpreted in terms of the upper and lower limits of the distribution.

(b) *Option for fitting c(s) and ls-g*(s) with 1 discrete species.* These options are particularly useful for fitting sedimentation velocity data for amyloid fibrils in the presence of small nonfibrillar components, such SAP or apoE. In this case, the parameters for the nonfibrillar component (the discrete species) can be determined in a separate experiment and used as fixed parameters in the fitting procedure.

(c) *Option for fitting for c(s) model with prior knowledge.* In general, fitting of experimental data to the c(s) Lamm equation (Eq. 7) involves estimating the diffusion coefficient for each *s*-value based on the weight average frictional ratio (23). However, when additional information is known about the protein system, it could be incorporated into the fitting procedure. For example, if the relationships between the sedimentation coefficient (*S*) and the molecular mass (*M*) is known, the diffusion coefficient (*D*) can be calculated for each *S* value using the Svedberg equation (23), and thus the relationship between *S* as a function *M* can be directly incorporated into the fitting procedure.

Figure 3 shows the fitted sedimentation coefficient distributions for the sedimentation velocity profiles shown in Fig. 2. The sedimentation velocity profiles in Fig. 2 were fitted to the c(s) model, with prior knowledge of the relationship between *S* and *M* for apoC-II, calculated using a model for worm-like chains for non-interacting fibrils. This relationship between sedimentation coefficient and fibril molecular weight has been described previously (17). The analysis of experimental data was performed with second-derivative Tikhonov–Phillips regularization (27) due to the broad size distribution of apoC-II fibrils. A regularization parameter of $p=0.68$ was used with 150 sedimentation coefficient increments. Time-independent noise was fitted and the sample meniscus position was fixed.

Initially, three discrete peaks are visible with modal sedimentation coefficients of approximately 10 S, 40 S, and 75 S (Fig. 3a). After 25 h incubation, the three peaks become more resolved and the amount of the 40 S and 75 S species increases at the expense of the small 10 S population (Fig. 3b). A characteristic bimodal sedimentation distribution was observed after 240 h incubation with a minor species of 40 S and major species of approximately 100 S (Fig. 3c). Based on previous studies, we attribute the 10 S species to short linear fibrils, 40 S to small closed loops (17, 32) and the major population of 75–100 S to long mature apoC-II fibrils. The area underneath the distributions is proportional to the amount of sedimenting solute and can be used to assess the fibrillar content of different samples over the incubation period.

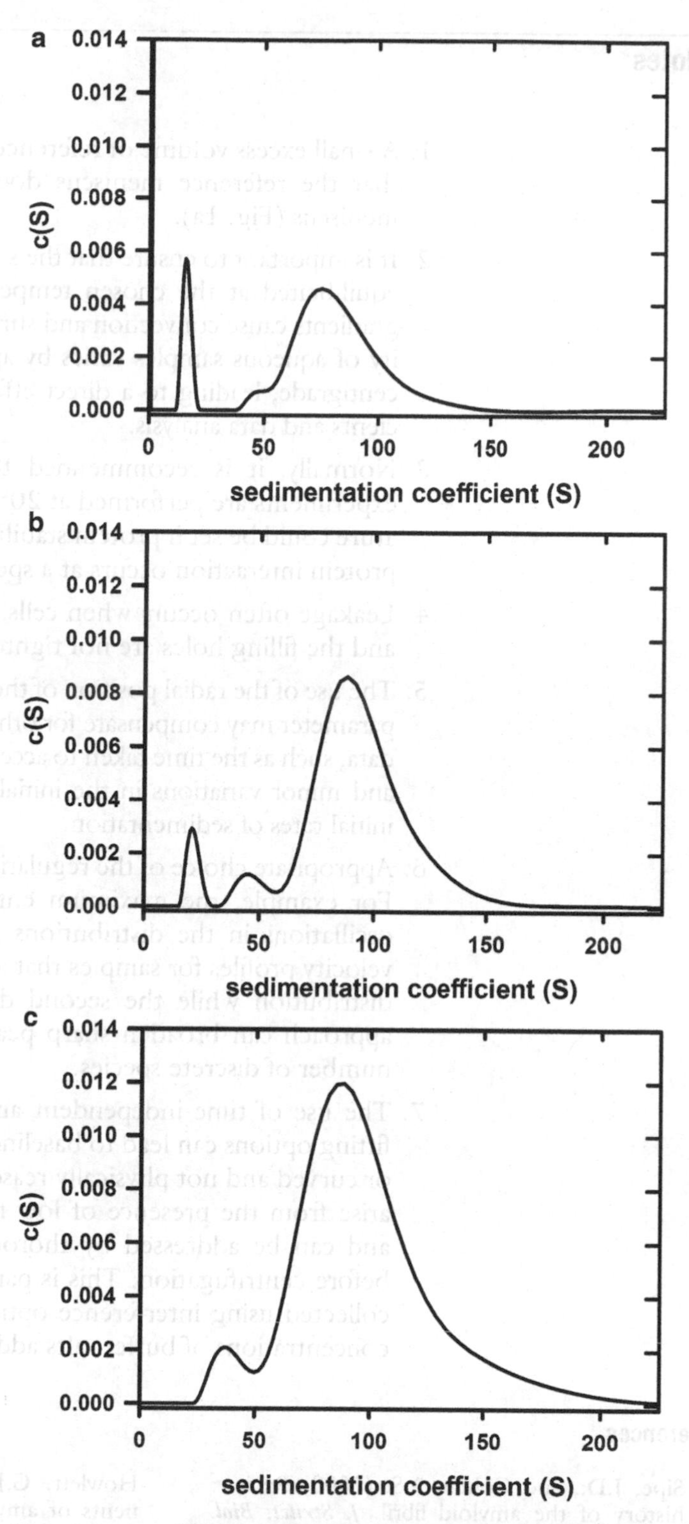

Fig. 3. Time-dependent continuous sedimentation coefficient distributions of amyloid fibrils as determined by c(s) model from the analysis of the sedimentation profiles shown in Fig. 2. Distributions for fibril formed after 10 h (**a**), 25 h (**b**) and 240 h (**c**) of incubation. Reproduced from Binger et al. (18) with permission from the publisher.

4. Notes

1. A small excess volume of reference sample is loaded to ensure that the reference meniscus does not obscure the sample meniscus (Fig. 1a).
2. It is important to ensure that the solution contents are properly equilibrated at the chosen temperature because temperature gradients cause convection and stirring. In addition, the viscosity of aqueous samples varies by approximately 3% per degree centigrade, leading to a direct effect on sedimentation coefficients and data analysis.
3. Normally, it is recommended that sedimentation velocity experiments are performed at 20°C. However, other temperature could be set if protein stability is a factor or the protein–protein interaction occurs at a specific temperature.
4. Leakage often occurs when cells are not tightened probably and the filling holes are not tightened sufficiently.
5. The use of the radial position of the sample meniscus as a fitting parameter may compensate for other uncertainties in fitting the data, such as the time taken to accelerate to the final rotor speed and minor variations in the initial temperature that affect the initial rates of sedimentation.
6. Appropriate choice of the regularization method is important. For example, the maximum entropy method can produce oscillations in the distributions when fitting sedimentation velocity profiles for samples that are expected to have a broad distribution while the second derivative Tikhonov–Phillips approach can broaden sharp peaks in samples containing a number of discrete species.
7. The use of time-independent and radial-independent noise fitting options can lead to baseline corrections that are sloped or curved and not physically reasonable. This difficulty could arise from the presence of low molecular weight impurities and can be addressed by thorough dialysis of the samples before centrifugation. This is particularly important for data collected using interference optics, where imbalances in the concentrations of buffer salts add to the measured signal.

References

1. Sipe, J.D., and Cohen, A.S. (2000) Review: history of the amyloid fibril. *J. Struct. Biol.* **130**, 88–98.
2. MacRaild, C.A., Stewart, C.R., Mok, Y.F., Gunzburg, M.J., Perugini, M.A., Lawrence, L.J., Tirtaatmadja, V., Cooper-White, J.J., and Howlett, G.J. (2004) Non-fibrillar components of amyloid deposits mediate the self-association and tangling of amyloid fibrils. *J. Biol. Chem.* **279**, 21038–21045.
3. Tennent, G.A., Lovat, L.B., and Pepys, M.B. (1995) Serum amyloid P component prevents

proteolysis of the amyloid fibrils of Alzheimer disease and systemic amyloidosis. *Proc. Natl. Acad. Sci. U.S.A.* **92**, 4299–4303.

4. Hawkins, P.N., and Pepys, M.B. (1995) Imaging amyloidosis with radiolabelled SAP. *Eur. J. Nucl. Med.* **22**, 595–599.
5. El-Agnaf, O.M., Nagala, S., Patel, B.P., and Austen, B.M. (2001) Non-fibrillar oligomeric species of the amyloid ABri peptide, implicated in familial British dementia, are more potent at inducing apoptotic cell death than protofibrils or mature fibrils. *J. Mol. Biol.* **310**, 157–168.
6. Kayed, R., Head, E., Thompson, J.L., McIntire, T.M., Milton, S.C., Cotman, C.W., and Glabe, C.G. (2003) Common structure of soluble amyloid oligomers implies common mechanism of pathogenesis. *Science* **300**, 486–489.
7. Reixach, N., Deechongkit, S., Jiang, X., Kelly, J.W., and Buxbaum, J.N. (2004) Tissue damage in the amyloidoses: Transthyretin monomers and nonnative oligomers are the major cytotoxic species in tissue culture. *Proc. Natl. Acad. Sci. U.S.A.* **101**, 2817–2822.
8. Bucciantini, M., Giannoni, E., Chiti, F., Baroni, F., Formigli, L., Zurdo, J., Taddei, N., Ramponi, G., Dobson, C.M., and Stefani, M. (2002) Inherent toxicity of aggregates implies a common mechanism for protein misfolding diseases. *Nature* **416**, 507–511.
9. Conway, K.A., Lee, S.J., Rochet, J.C., Ding, T.T., Williamson, R.E., and Lansbury, P.T., Jr. (2000) Acceleration of oligomerization, not fibrillization, is a shared property of both alpha-synuclein mutations linked to early-onset Parkinson's disease: implications for pathogenesis and therapy. *Proc. Natl. Acad. Sci. U.S.A.* 97, 571–576.
10. Lambert, M.P., Barlow, A.K., Chromy, B.A., Edwards, C., Freed, R., Liosatos, M., Morgan, T.E., Rozovsky, I., Trommer, B., Viola, K.L., Wals, P., Zhang, C., Finch, C.E., Krafft, G.A., and Klein, W.L. (1998) Diffusible, nonfibrillar ligands derived from Abeta1-42 are potent central nervous system neurotoxins. *Proc. Natl. Acad. Sci. U.S.A.* **95**, 6448–6453.
11. Caughey, B., and Lansbury, P.T. (2003) Protofibrils, pores, fibrils, and neurodegeneration: separating the responsible protein aggregates from the innocent bystanders. *Annu. Rev. Neurosci.* **26**, 267–298.
12. Hammarstrom, P., Jiang, X., Deechongkit, S., and Kelly, J.W. (2001) Anion shielding of electrostatic repulsions in transthyretin modulates stability and amyloidosis: insight into the chaotrope unfolding dichotomy. *Biochemistry* **40**, 11453–11459.
13. Lashuel, H.A., Lai, Z., and Kelly, J.W. (1998) Characterization of the transthyretin acid denaturation pathways by analytical ultracentrifugation: implications for wild-type, V30M, and L55P amyloid fibril formation. *Biochemistry* **37**, 17851–17864.
14. Lashuel, H.A., Hartley, D.M., Balakhaneh, D., Aggarwal, A., Teichberg, S., and Callaway, D.J. (2002) New class of inhibitors of amyloid-beta fibril formation. Implications for the mechanism of pathogenesis in Alzheimer's disease. *J. Biol. Chem.* **277**, 42881–42890.
15. Lashuel, H.A., Petre, B.M., Wall, J., Simon, M., Nowak, R.J., Walz, T., and Lansbury, P.T., Jr. (2002) Alpha-synuclein, especially the Parkinson's disease-associated mutants, forms pore-like annular and tubular protofibrils. *J. Mol. Biol.* **322**, 1089–1102.
16. Smith, A.M., Jahn, T.R., Ashcroft, A.E., and Radford, S.E. (2006) Direct observation of oligomeric species formed in the early stages of amyloid fibril formation using electrospray ionisation mass spectrometry. *J. Mol. Biol.* **364**, 9–19.
17. MacRaild, C.A., Hatters, D.M., Lawrence, L.J., and Howlett, G.J. (2003) Sedimentation velocity analysis of flexible macromolecules: self-association and tangling of amyloid fibrils. *Biophys. J.* **84**, 2562–2569.
18. Binger, K.J., Pham, C.L., Wilson, L.M., Bailey, M.F., Lawrence, L.J., Schuck, P., and Howlett, G.J. (2008) Apolipoprotein C-II amyloid fibrils assemble via a reversible pathway that includes fibril breaking and rejoining. *J. Mol. Biol.* **376**, 1116–1129.
19. Fujita, H. (1962) Mathematical theory of sedimentation analysis (New York: Academic Press).
20. Holloday, L. (1979) An approximate solution of the Lamm equation. *Biophys. Chem.* **10**, 187–190.
21. Philo, J.S. (1997) An improved function for fitting sedimentation velocity data for low-molecular-weight solutes. *Biophys. J.* **72**, 435–444.
22. VanHolde, K., and Weischet, W. (1978) Boundary analysis of sedimentation velocity experiments with monodisperse and paucidisperse solutes. *Biopolymers* **17**, 1387–1403.
23. Schuck, P. (2000) Size-distribution analysis of macromolecules by sedimentation velocity ultracentrifugation and Lamm equation modeling. *Biophys. J.* **78**, 1606–1619.
24. Demeler, B., and Saber, H. (1998) Determination of molecular parameters by fitting sedimentation data to finite-element solutions of the Lamm equation. *Biophys. J.* **74**, 444–454.
25. Schuck, P. (1998) Sedimentation analysis of noninteracting and self-associating solutes

using numerical solutions to the Lamm equation. *Biophys. J.* **75**, 1503–1512.

26. Devlin, G.L., Chow, M.K., Howlett, G.J., and Bottomley, S.P. (2002) Acid Denaturation of alpha1-antitrypsin: characterization of a novel mechanism of serpin polymerization. *J. Mol. Biol.* **324**, 859–870.
27. Schuck, P., and Rossmanith, P. (2000) Determination of the sedimentation coefficient distribution by least-squares boundary modeling. *Biopolymers* **54**, 328–341.
28. Stafford, W.F., 3rd (1992) Boundary analysis in sedimentation transport experiments: a procedure for obtaining sedimentation coefficient distributions using the time derivative of the concentration profile. *Anal. Biochem.* **203**, 295–301.
29. Philo, J.S. (2000) A method for directly fitting the time derivative of sedimentation velocity data and an alternative algorithm for calculating sedimentation coefficient distribution functions. *Anal. Biochem.* **279**, 151–163.
30. Schuck, P., and Demeler, B. (1999) Direct sedimentation analysis of interference optical data in analytical ultracentrifugation. *Biophys. J.* **76**, 2288–2296.
31. Schuck, P., Perugini, M.A., Gonzales, N.R., Howlett, G.J., and Schubert, D. (2002) Size-distribution analysis of proteins by analytical ultracentrifugation: strategies and application to model systems. *Biophys. J.* **82**, 1096–1111.
32. Hatters, D.M., MacRaild, C.A., Daniels, R., Gosal, W.S., Thomson, N.H., Jones, J.A., Davis, J.J., MacPhee, C.E., Dobson, C.M., and Howlett, G.J. (2003) The circularization of amyloid fibrils formed by apolipoprotein C-II. *Biophys. J.* **85**, 3979–3990.

Chapter 13

Transmission Electron Microscopy of Amyloid Fibrils

Sally L. Gras, Lynne J. Waddington, and Kenneth N. Goldie

Abstract

Transmission Electron Microscopy of negatively stained and cryo-prepared specimens allows amyloid fibrils to be visualised at high resolution in a dried or a hydrated state, and is an essential method for characterising the morphology of fibrils and pre-fibrillar species. We outline the key steps involved in the preparation and observation of samples using negative staining and cryo-electron preservation. We also discuss methods to measure fibril characteristics, such as fibril width, from electron micrographs.

Key words: TEM, Cryo-EM, Negative staining

1. Introduction

Transmission Electron Microscopy (TEM) is a useful technique for assessing the *in vitro* formation of amyloid fibrils from proteins or peptides. TEM allows researchers to see structural features at the nanometre scale that cannot be visualised by eye or light microscopy. This technique can resolve features as small as 1–2 nm and provide images of possible early aggregates, circular species, protofilaments and mature fibrils (1, 2). It can also provide evidence of other protein mis-folding pathways, such as the formation of amorphous aggregates (3). Cryo-electron microscopy (cryo-EM), a specialised TEM technique where samples are rapidly frozen preserving their more native hydrated state, can also be used to characterise mis-folded proteins (4, 5) and has recently been applied to examine fibrillar structures formed on the surface of large unilamellar vesicles (6).

TEM images can be used for the qualitative comparison of features, such as the twists in ribbon-like fibrils, curvature of fibrils and the smoothness of their surface (7). TEM images can also be

Andrew F. Hill et al. (eds.), *Protein Folding, Misfolding, and Disease: Methods and Protocols*,
Methods in Molecular Biology, vol. 752, DOI 10.1007/978-1-60327-223-0_13, © Springer Science+Business Media, LLC 2011

used to generate quantitative data, including the length of early aggregates and seeds, width of fibrils, number of protofilaments and the periodicity of fibril twists (1). One of the attractions of TEM analysis is that it can be performed relatively quickly, allowing researchers to rapidly assess whether fibril formation has occurred and quickly determine how a particular batch of fibrils is tracking relative to previous experiments with that protein or peptide species.

While TEM can confirm the presence of a fibril-like morphology, it cannot confirm the presence of cross-beta sheets, and other techniques may be required in combination with TEM (8) to determine whether the fibrillar structure has the characteristics of an amyloid fibril or whether the proteins and peptides are arranged in another configuration. TEM provides greater resolution than other microscopic techniques, such as confocal microscopy and scanning electron microscopy; it also avoids the need for metal coating often required for the latter technique. Atomic Force Microscopy (AFM) is complementary to TEM and the two techniques are often combined to fully characterise a fibril-forming system. AFM can be used to observe hydrated or dehydrated samples under atmospheric conditions (c.f. hydrated and dehydrated samples in a vacuum for cryo-EM and TEM, respectively). AFM can also provide height and topographic information not available with TEM. However, an advantage of cryo-EM, compared to AFM, is that samples need not be adsorbed on the surface prior to freezing and imaging.

In the following pages, we explain how to prepare samples for TEM using negative staining, how to analyse samples by TEM, how to measure the key features in a TEM image and how to perform cryo-EM. TEM can also be used to examine thin cross-sections of amyloid fibrils (9) or amyloid fibrils occurring *in situ* within cellular tissues that have been embedded in resin (10). This method can be used to investigate the significance and function of amyloid structures in a cellular context and can also be employed to determine the width of a fibril and the number of protofilaments and assess whether a fibril has an electron lucent or electron-dense centre (10). These types of images can also be averaged to produce a reconstruction either in cross-section or from the side (11). Other techniques include immuno-labelling, which can be used to test whether particular residues are exposed on the surface of the fibril (12, 13) and metal shadowing to determine the handedness of fibrils (14). These latter methods are beyond the scope of this chapter and readers are referred to a range of specialised references for further details on these techniques (15, 16).

1.1. Negative Staining for TEM

Negative staining is a relatively quick method for the preparation of samples for imaging by TEM. It is suitable for amyloid fibrils in solution, small particles such as viruses, cell components such as ribosomes, single molecules or complexes of molecular weight above approximately 100 kDa and various

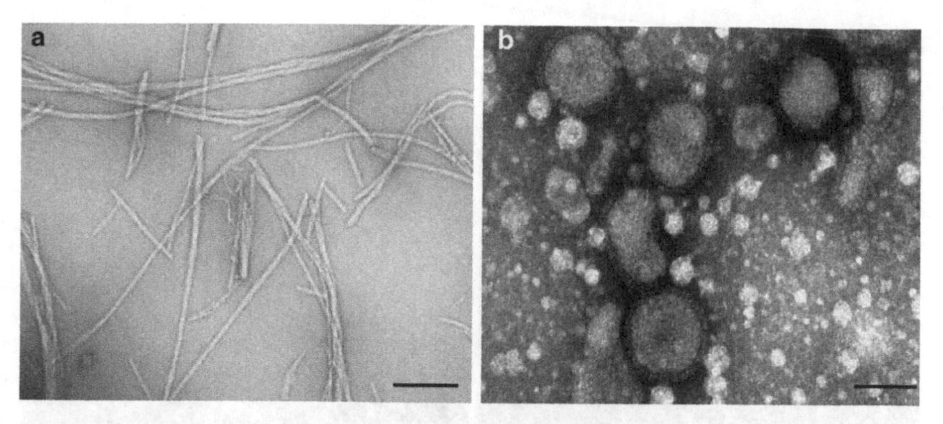

Fig. 1. Negatively stained TEM images. (**a**) Alzheimer's Aβ fibrils and (**b**) Avian Influenza Virus. Note the light appearance of the fibrils against the darker stain background and the partial penetration of the stain into the virus particles, revealing some internal structure, as well as the "fringe" of neuraminidase and haemagglutinin molecules around the edge of each virus. The scale bars are 200 nm and 100 nm in length, respectively.

nanoparticles as illustrated in Fig. 1. Negative staining is not suitable for larger objects, such as whole cells, and most bacteria are too large to be successfully imaged by this technique.

Negative staining typically generates the sample with good contrast and well-preserved morphology, as the stain not only provides contrast but also protects the specimen from radiation damage. The stain – usually a metallic salt such as uranyl acetate or phosphotungstic acid – forms a coating over the sample, embedding it in a layer of radiation-stable, electron-dense material. The sample can partially exclude stain due to its higher topology compared to the stain layer, which remains on the surface of the specimen support film. Stain salts also tend to build up at the edges, where the sample adheres to the surface, resulting in an increased stain density in these regions. The electron beam passes through the sample but is absorbed by the metallic stain, hence the sample appears light and the surrounding stain appears dark. The stain also penetrates into holes and crevices in the sample, revealing details of the substructure.

Details of the precise mechanism of negative staining are discussed at length in electron microscopy texts and in the literature (15, 17). One aspect which becomes clear is that different negative stains reveal different aspects of sample morphology and indeed some negative stains are unsuccessful with certain samples. The most commonly used negative stains are uranyl acetate and phosphotungstic acid as mentioned above. A more extensive list of negative stains and their applications is given in (15). The method described in this chapter uses uranyl acetate.

1.2. Cryo-EM

Cryo-EM, sometimes called electron cryomicroscopy, was first described over 30 years ago (18, 19) as a means to reduce radiation damage that occurs when a sample is imaged in an electron microscope.

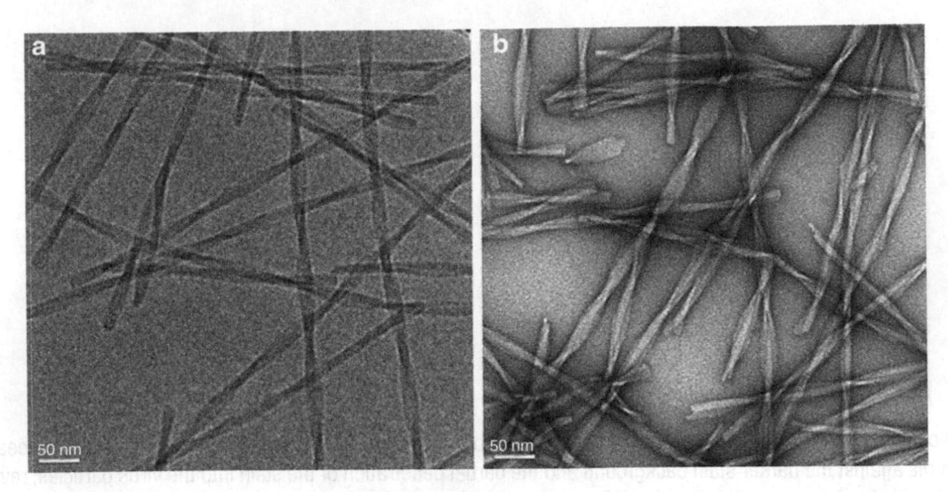

Fig. 2. Comparison of a Cryo-EM image (**a**) and negatively stained TEM image (**b**) of an amyloid fibril constructed from a TTR1-based synthetic peptide. The scale bar in each image is 50 nm in length.

Cryo-EM generally refers to the imaging of samples in the TEM at cryogenic temperatures and is most commonly applied in the life sciences to isolated macromolecular assemblies, such as amyloid fibrils. It is recognised as the most close-to-native preparation method currently available for the TEM of such samples. Figure 2 provides a comparison of an amyloid fibril preparation imaged by Cryo-EM and by using the negative stain procedure outlined in the above Subheading 1.1.

Cryo-EM involves the rapid freezing of the sample, commonly employing plunge freezing of the specimen into either liquid ethane or liquid propane. Freezing rates of around 10,000 K/s produce an amorphous (vitreous) ice layer (~100 nm thick) from the solute buffer that essentially "embeds" the sample. Such a preparation regime has numerous advantages over negative staining techniques. For example, the sample is not exposed to salt or low pH as occurs with a negative stain. The sample remains hydrated and because it does not undergo dehydration, retains its 3-dimensional shape within the vitreous ice layer. In addition, as the sample is suspended within a hole in a carbon film, it does not come into direct contact with the carbon surface; so any charge interactions that may occur between sample and support film are negated.

1.3. Cryo-Negative Stain

The recently developed technique of cryo-negative staining allows the possibility to improve contrast for the study of plunge-frozen biological samples with the electron microscope. Samples are observed at low temperature, in the vitrified state, but in the presence of a stain (usually ammonium molybdate) (20).

The advantage of incorporating a negative stain with cryo-prepared specimens is that contrast can be greatly enhanced with a signal-to-noise ratio significantly higher than that for unstained samples. It has been reported (21) that beam damage can be considerably reduced using this technique, as shown by the comparison of multiple-exposure series of both stained and unstained samples. However, this technique cannot be employed with all samples as some specimens, for example microtubules, show sensitivity when exposed to the high salt content of the stain and consequently exhibit poor preservation.

2. Materials

The experiments described here assume that users have access to standard laboratory equipment, such as pipettes.

2.1. Negative Staining for Transmission Electron Microscopy

1. Carbon-coated grids, copper, and 300 mesh (ProSciTech, Thuringowa, Queensland, Australia, see Note 1).
2. Glass microscope slide.
3. Filter paper cut into strips or wedges. A fast filter paper, such as Whatman 541 works well.
4. Fine-tipped forceps for handling grids (ProSciTech, see Note 2).
5. Distilled or water purified to a resistivity < 18.2 Ω (e.g. from a MilliQ unit).
6. Uranyl Acetate solution: 2% (w/v) in distilled water (see Notes 3 and 4 on safe handling).
7. 5–10 μl pipette tips.
8. Grid box or glass Petri dish lined with filter paper for storing prepared grids.
9. Dried silicon for longer term storage of the grid box containing grids.
10. Lint-free tissue (e.g. Kim wipes).
11. Glow-discharge unit (homebuilt or commercial, e.g. Emitech, (Quorum Technologies Ltd, Kent, UK)).

2.2. Transmission Electron Microscopy

Most universities have a TEM suitable for biological imaging in the biological, medical or physics department. Basic requirements include the microscope and a camera unit. New instruments are likely to include a CCD detector, whereas older instruments may still be fitted with a camera and film. This chapter assumes that the user has access to a newer instrument with CCD detector. Other requirements are:

1. Samples that have been negatively stained (see Subheading 2.1).
2. Fine-tipped forceps for handling grids (see Subheading 2.1).
3. Thumb drive or other means of transferring images from the computer.

2.3. Image Analysis

1. ImageJ software. This can be downloaded for free from the NIH Web site, at http://rsb.info.nih.gov/ij.
2. Microsoft Excel or equivalent software for manipulating and plotting data.

2.4. Cryo-Electron Microscopy

1. Ammonium molybdate stain (Proscitech) if negative staining is to be applied. This is typically applied as a saturated solution. Prepare a slurry of ~1.0 g of ammonium molybdate tetrahydrate in 0.9 ml of water (not deionised). Shake the sample and add 0.1 ml of 10 M NaOH and shake again. A more detailed protocol for the preparation of this stain can be found in reference (22).
2. Supply of liquid nitrogen which is free from frost contamination.
3. Plexiglass face shield or safety glasses: In addition to protecting your eyes, the full-face shield has the advantage that it also helps direct your breath away from the sample, preventing frost contamination on the specimen surface.
4. High-purity gaseous ethane (see Note 5): This should be fitted with a gas regulator valve fitted with a needle valve that can be used to adjust and shut off the flow of ethane. The outlet of the regulator should also be fitted with a length of 6 mm diameter silicon tubing. A plastic pipette tip is then inserted to the end of the tubing. The pipette tip is cut at a 45° angle, so the gas flow is not blocked if the tip hits the bottom of the ethane cup.
5. Grids – usually 200- or 300-mesh copper grids which have an additional fenestrated support film (Quantifoil Micro Tools GmbH, Jena, Germany).
6. Glow-discharge unit (see Subheading 2.1 above).
7. Rapid plunge-freezing device: This may be an automated commercial device, for example, a Vitrobot (FEI Company, Eindhoven, the Netherlands), or a home-made device produced in a workshop. Some units may house a preparation chamber which allows temperature and humidity control. The entire device should be placed into an extraction hood to remove gaseous ethane. The surrounding area should be spark-free to prevent the ignition of ethane. An insulated container sits at the base of the plunger as a nitrogen reservoir, into which a small container sits for condensing ethane. Liquid ethane forms as the freezing medium.
8. Qualitative No. 1 filter paper (Whatman Cat. No. 1001055) for blotting.
9. Cryo-storage grid boxes: These can be made in a workshop by cutting up grid boxes (Agar Scientific, Cat. No. G276A) into 15 mm squares so that each square contains four original

grid holes arranged symmetrically. Circular rotatable lids of 1 mm-thick clear plastic are fitted with a nylon screw drilled at the centre of the four holes. A small opening is cut into the plastic lid to allow access for the removal of a single grid while keeping the other holes covered.

10. 50 ml plastic conical tubes (BD Biosciences, NJ, USA) for the storage of grid boxes. Two 10 mm holes should be made in the lid of the tube to allow nitrogen flow during storage and to allow tube retrieval with long forceps from the storage Dewar.
11. Long-term liquid nitrogen storage Dewar (MVE, Germany) fitted with separate canisters for storing frozen grids in grid boxes housed inside plastic conical tubes.
12. Portable Dewar (KGW-Isotherm, Karlsruher, Germany): This Dewar is used for grid transport between the main Dewar and microscope, once the grids with sample are frozen and placed in the grid box.
13. Large 200 mm tweezers to manipulate grid boxes and 300 mm tweezers to remove plastic conical tubes from the long-term storage Dewar.
14. Small screw drivers to allow the opening and closing of grid box lids by loosening and tightening their nylon screw.
15. Fine forceps (e.g. number 5 Dumont, Switzerland) for grid handling. Pipette tips for applying samples to grids.
16. Cryo-capable transmission electron microscope equipped with nitrogen-cooled anti-contaminator system (cryo-blades or a cryo-box).
17. Computer software integrated with the microscope software to allow image acquisition using low-dose methods.
18. Cryo-holder (Gatan, Inc., Pleasanton, CA, USA) for inserting frozen specimens into the TEM and maintaining the temperature at approximately –170 °C.
19. Cryo-transfer station, necessary for loading plunge-frozen grids into the Gatan cryo-holder at liquid nitrogen temperature for subsequent loading into the microscope.

3. Methods

3.1. Negative Staining for Transmission Electron Microscopy

1. Samples containing aggregated protein for TEM analysis should be in a buffer, such as Tris-Buffered Saline, or in water at a concentration of 0.01–0.5 mg/ml. It is usually necessary to try a range of sample dilutions in order to obtain a field of view in which particles are neither too sparse nor overlapping. Dilution should occur immediately prior to the sample preparation to avoid changes in the sample (see Notes 6 and 7).

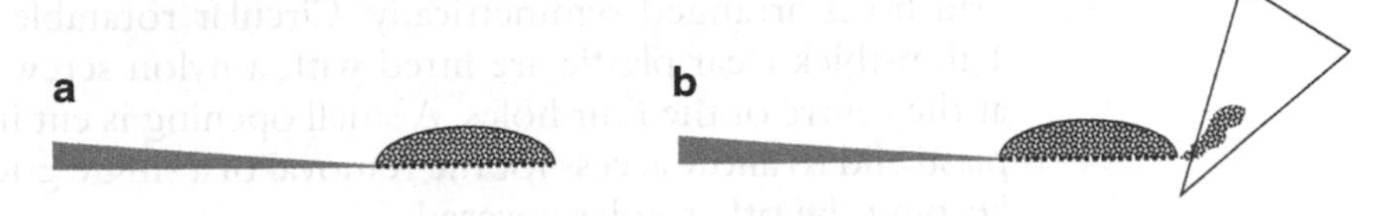

Fig. 3. Experimental set-up for negative staining. (**a**) A carbon-coated grid is mounted in a pair of forceps and (**b**) a wedge of filter paper is used to blot the liquid added to the grid by gently touching the side of the grid.

2. Using forceps to handle the grids, place two to three carbon-coated grids carbon side up on a glass microscope slide. The carbon side is the dull side. Transfer the slide into the Glow-discharge unit. Operation of the Glow-discharge unit varies depending on the model available. Glow-discharge the grids for 15 s in 0.1 Torr of Nitrogen (see Note 8).
3. Mount a grid carbon-side-up in a pair of forceps, holding the grid only by the rim to avoid damage. Place the forceps and grid in a stable position on the lab bench as shown in Fig. 3a. If several grids are to be stained, it is most efficient to set up a series of tweezers in parallel.
4. Carefully apply 4–5 μl of sample to the grid using a pipette. Care must be taken so that the pipette tip does not touch the grid surface directly as this can damage the grid. Allow the sample to settle for 1 min, observing continuously to ensure that surface tension does not cause the sample to run up the inside of the forceps. If this occurs, use a smaller aliquot of sample and rinse the forceps thoroughly.
5. Next, use a wedge of filter paper to wick the sample off the grid by touching the filter paper at right angles to the edge of the grid, as shown in Fig. 3b.
6. Quickly, without letting the sample dry, apply a similar aliquot of distilled water with a fresh pipette tip and wick off as before (see Note 9).
7. Quickly apply a similar aliquot of uranyl acetate. Allow this solution to stain for 30 s and then wick off with filter paper and allow to air dry.
8. The grid is now ready to transfer to the electron microscope or for storage in a grid box or covered Petri dish (see Note 10). Samples are typically stored with the sample surface towards the left in the grid box or the sample surface on top in a Petri dish.
9. The tweezers should then be rinsed well with water and blotted with a lint-free tissue to prevent contamination of the next experiment, as the tips come into direct contact with the sample.

3.2. TEM Imaging and Data Acquisition

It is beyond the scope of this chapter to describe in detail how to operate an electron microscope, so we assume that users have had training in operation of their TEM and are able to load the sample correctly and perform the basic alignment, including correcting eucentric height, astigmatism and focus. We shall also assume that the microscope is either correctly set up for CCD (for example, intensity level and gain calibration) imaging or, in the case of older microscopes, for film (intensity level, exposure, sensitivity).

3.2.1. Magnification

Negatively stained samples are limited in resolution to about 20–25 Å, this being influenced by the grain size of the stain used and its interaction with the sample. The ideal magnification, therefore, is the minimum magnification required to achieve this resolution. This varies according to the size of the CCD detector and the number of pixels, as a rough guide magnifications between 20,000× and 80,000× are useful. Use of too high a magnification does not provide any higher resolution and increases the amount of radiation damage suffered by the sample.

3.2.2. Choice of Area to Image, Degree of Staining, and Depth of Stain

Amyloid fibrils typically distribute unevenly on a grid, leading to some areas with very few fibrils and others with dense clusters or layers of fibrils. The dense areas are usually very darkly stained and little detail of fibril morphology is visible. One should, therefore, look at low magnification – about 5,000× – for areas of low-fibril density with even, uniform staining and then image these areas at the desired higher magnification. Figure 4 illustrates a typical sample of amyloid fibrils viewed at lower and higher magnifications.

3.2.3. Focus or Underfocus

Negatively stained samples are intrinsically high in contrast and can, therefore, be imaged close to focus. When an image is perfectly focused, it has minimum contrast and the contrast increases as

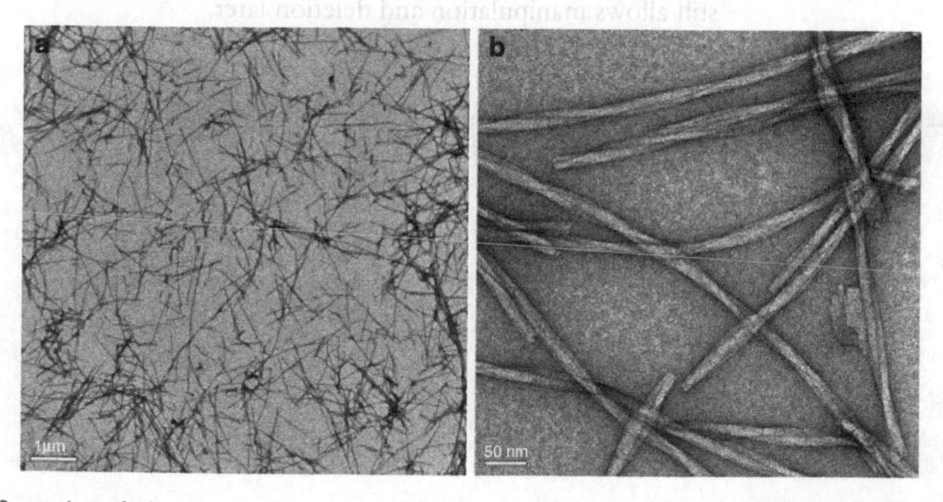

Fig. 4. Comparison of a low (**a**) and high (**b**) magnification image of negatively stained amyloid fibrils constructed from a TTR1-based synthetic peptide, illustrating the typical conditions used to look for a suitable area to image on the grid and to then examine the features of the amyloid fibril sample. The scale bars are 1 μm and 50 nm in length, respectively.

the image is defocused. Negatively stained samples do not have to be defocused greatly compared to cryo-samples, which are often imaged at defocus levels of the order of 2 μm.

3.2.4. Artefacts

In all electron microscopy studies, it is important to be aware of the possibility that images, which do not accurately represent the sample, will be produced. Some of these artefacts are a consequence of specimen preparation technique; other artefacts are a result of contaminants in the sample. Contamination is particularly relevant to amyloid studies. For example, the fibril-like structures shown in Fig. 5 were found in filtered distilled water (Fig. 5a) and phosphate-buffered saline (Fig. 5b). This illustrates the importance of examining more than one grid for each sample.

3.2.5. Data Export

Digital images from CCD cameras are generated in a variety of formats. The most common format in electron microscopy is Digital Micrograph's .dm3 and the .tif format of Olympus's AnalySiS software. Note that the .dm3 proprietary format and some other formats, including tiff (especially those saved in 16 bit), are not currently readable by some image-processing software packages; however, Image J (see Subheading 3.3 below) has plug-ins that support the most common TEM image formats (including .dm3). TEM files can also be exported from the TEM software in a regular tiff form, in the case of Digital Micrograph, or as a .jpg in the case of AnalySiS. Be aware, however, that .jpg-formatted images deteriorate in quality when subjected to successive rounds of image processing. A further consideration to note is that at the stage of saving an image, you usually have the option to include a scale bar (see Note 11). Saving in some formats (e.g. tiff) then results in the scale bar being integrated into the image and cannot be removed later, while saving as .dm3, for example, still allows manipulation and deletion later.

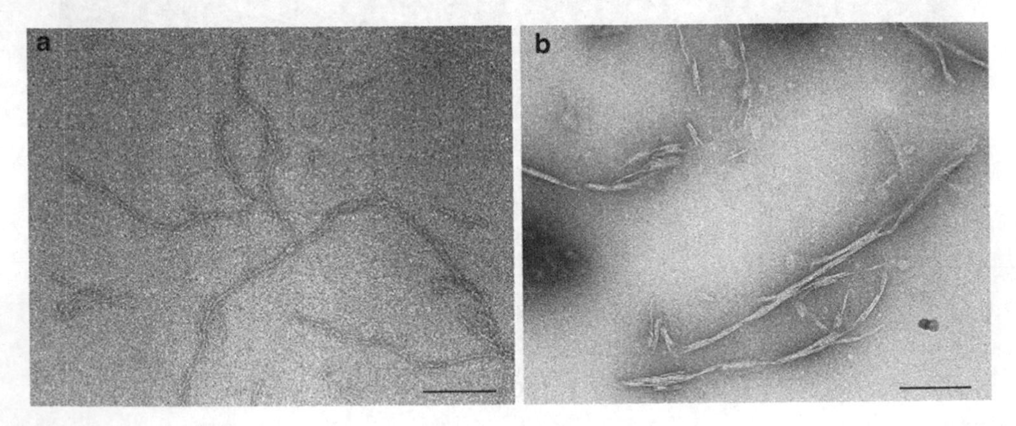

Fig. 5. Typical artefacts observed by TEM which can originate from the buffer or the grid surface. (**a**) A water sample and (**b**) a phosphate-buffered saline sample. The scale bar in each image is 100 nm in length.

3.3. Image Analysis

Image analysis can be performed once a TEM image has been saved and transferred from the microscope. The software package, Image J, can be used for image analysis, although a number of other packages are also available (including software that may be present on the TEM microscope). Image J may be applied to measure features, such as fibril length or thickness as follows:

1. Instal ImageJ following the instructions from the NIH Web site (see Subheading 2.3).
2. Open ImageJ.
3. Open the TEM image (see Subheading 3.2 above and see Note 12).
4. Using the straight line selection drawing tool, draw a line that spans the scale bar (see Note 11).
5. Open the set scale function under the analyse toolbar and add the known length and unit.
6. Measure the desired feature on the fibril or other aggregate using the straight line drawing tool and apply the measure tool under the analyse toolbar to bring up a window with the length in pixels and the unit of your scale bar.
7. Save the measurements and then repeat step 6 to get a number of measurements (see Notes 13 and 14).
8. Export the measurement data by saving the file as measurements. This file type can then be opened and analysed in Excel.
9. When reporting this type of data, it is preferable to quote the mean of the data ± the standard deviation and the number of fibrils measured.

3.4. Sample Preparation for Cryo-EM

When performing rapid plunge freezing, the goal is to obtain an evenly distributed specimen, quick-frozen in a thin homogeneous layer of vitrified ice. Ideally, the sample is free of frost contamination and should be at a reasonable concentration so as not to be either too concentrated (causing too much overlap or clustering of the sample) or too dilute, necessitating excessive searching across the grid (see Note 15). The blotting regime is, therefore, critical. The amount of blotting required can be sample- and/or buffer-dependent, so might require some adaptation for each specimen. It is important to retain consistent parameters but because in some cases even sub-second differences in blotting time can cause variations in the ice thickness, differences between grids are unavoidable. In order to counter this problem, several grids should be prepared during each plunging session.

1. Prepare the working area so that all necessary tools and other materials are in close proximity to the plunging device, as shown in Fig. 6. Ensure adequate lighting and check that

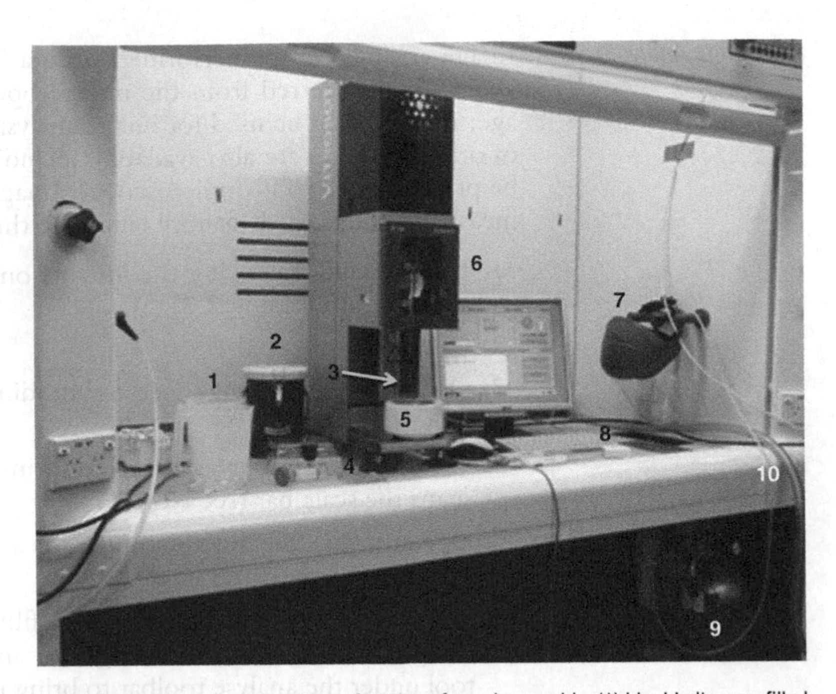

Fig. 6. Equipment set-up for plunge freezing of specimen grids. (1) Liquid nitrogen-filled plastic beakers for topping up nitrogen in nitrogen bath, (2) Frozen sample transport Dewar. (3) Grid held by a pair of forceps fixed to the plunging specimen rod. (4) Plastic grid boxes and tubes for storage of frozen grids. (5) Insulated container containing a metal bath for liquid nitrogen, which in turn surrounds a small container for condensing ethane. (6) Temperature and humidity-controlled environmental blotting chamber. Note white filter paper pads. The computer and keyboard to the right are used for controlling the blotting process. (7) Full-face shield. (8) Box containing blotting paper, pipettes for sample application and tweezers for grid box handling under nitrogen. (9) Gaseous ethane bottle. (10) Gas regulator valve with attached silicon tubing and pipette tip for delivering ethane.

there are no ignition sources in the area. The plunger should be placed under a fume hood if possible to allow the extraction of ethane vapours. Try to minimise frost contamination in the cryogens by avoiding breathing onto the liquid surfaces and grid. Pre-cool all tools that come into contact with the specimen grid and grid box to avoid heating the sample above its devitrification temperature once frozen (approximately −140 °C).

2. Prepare your purified sample for application (see Note 16).
3. Switch on the plunging device if it is automated and set the parameters (these have to be tested and adjusted for a particular sample/buffer).
4. Wearing eye protection or a full-face shield (which helps to minimise breathing onto sample), place a grid box into the base of the nitrogen reservoir and then fill the container with liquid nitrogen. Allow the temperature to come down

and stabilise. Note: Always avoid cryogen contact with the skin and eyes and be aware that ethane can cause severe cold burns and is highly flammable.

5. Hold the pipette tip into the bottom of the ethane pot and slowly open the ethane supply and needle-regulating valve to slowly condense the ethane to a liquid. Carefully fill the ethane container to the top.
6. Place a grid (normally freshly glow-discharged so that the grid surface is hydrophilic) edge within forceps that fit the plunging device.
7. If the grid requires water washing (e.g. to remove residual salt), apply 3–5 μl of sample with the grid in the tweezers sitting face-up on the bench, wait 30–60 s, then blot and wash 2× in distilled water, blotting between each wash. Avoid the sample drying out between steps. Then, fix the tweezers to the plunging arm of the plunge device. If the sample does not require washes, with an automated plunger you can directly apply the sample to the grid within the sample chamber.
8. After sufficient adsorption time (30–60 s), activate the automated blotting and plunging process. Cryo-negative staining requires an additional step prior to blotting. This procedure is given in detail in reference (22).
9. Once plunged, the rapid-frozen sample is then carefully removed from the ethane and, keeping it in cold nitrogen vapour, quickly transferred below the liquid nitrogen level in the surrounding nitrogen container and placed in the pre-cooled grid box (see Note 17). The grid can either be transferred to a long-term liquid nitrogen container for storage or directly into the microscope. A plastic conical tube is used for the placement of grid boxes (several boxes can be placed in one labelled tube) and is then inserted into racks within a suitable nitrogen storage Dewar.

3.5. Grid Transfer to the Microscope for Cryo-EM

1. Prepare the microscope for cryo-EM by filling nitrogen into the anti-contaminator device that is integrated into the microscope.
2. Remove the cryo-holder from the vacuum pump stand (it is recommended to continually pump the holder, when not in the microscope, to maintain a good vacuum in the holder Dewar for optimal thermal insulation).
3. Carefully insert the cryo-holder into the transfer station. Place on a bench adjacent to the microscope for loading.
4. Place a plastic conical tube into a small nitrogen Dewar and fill both the tube and Dewar with liquid nitrogen.

5. Take the grid boxes housed within a conical tube from the storage Dewar and transfer the required grid boxes into the nitrogen-filled transfer tube. Replace any grid boxes not to be viewed back into the storage Dewar for safe keeping. Take the Dewar holding the grids to the loading station for observation.
6. Cool tweezers and other tools that come into contact with the grid or grid boxes.
7. Pour liquid nitrogen to cool the loading station and cryo-holder Dewar. Once stable at liquid nitrogen temperature (at least –160 °C, measured with a temperature controller), transfer the grid boxes into the nitrogen contained in the loading reservoir.
8. Transfer the grid into the holder and fix the grid firmly with the clamping mechanism (usually some form of clip ring); close the cryo-shutter to avoid sample warming and to protect the grid from water vapour during transfer.
9. Move the loading station and holder to the microscope and pre-pump the vacuum lines if available on the microscope. Carefully remove the holder from the loading station and insert it into the microscope stage for pre-evacuation (see Note 18). Once pre-pumping has completed, the holder can be inserted all the way into the column, gripping the holder while simultaneously resetting the stage to its 0 tilt position.
10. Allow the temperature of the holder to equilibrate and stabilise, which take around 20 min. Most holders reach a constant operating temperature of approximately –170 °C.

3.6. Cryo-EM Imaging and Data Acquisition

In order to minimise specimen damage by the electron beam during image acquisition in the microscope, it is necessary to apply low-dose imaging techniques, which require special software for beam blanking and microscope control. Additional low-dose alignments and parameter settings on the microscope are required to optimise the operation and protect the sample from excessive electron beam radiation.

The grid is usually screened at a low magnification for the selection of a suitable position (with good ice thickness and sample distribution). Once a promising area is located, an image shift is applied which moves the beam away from the position of interest and sets this area at a higher magnification so that the sample can be carefully focussed. (Usually, a degree of defocus is applied to enhance image contrast). Then, the software shifts back to the area of interest, and a single image is taken at a predetermined exposure magnification. Successive images are then acquired from new areas across the grid.

3.7. Cryo-EM Image Processing

As a result of the inherent low contrast obtained with cryo-EM images, it is often advantageous to apply some form of image processing. Post-acquisition imaging treatment methods can be indispensable for enhancing the signal to noise and improving attainable resolution. These often involve various averaging techniques which can be applied to specimens that comprise features which exist as multiple copies, as is the case, for example, with some filamentous structures which have a repeating helical symmetry. The routines and methods of image processing are complex and out of the scope of this article. Several excellent books and review articles exist in the literature and readers are encouraged to investigate these further.

4. Notes

1. Copper grids are the cheapest/and are suitable for basic negative staining experiments. They can be purchased pre-coated with carbon film. 300 mesh is recommended for most purposes. Larger mesh sizes give greater surface area unobstructed by grid bars but the carbon film becomes increasingly fragile and hard to handle with a larger mesh.
2. When using nickel grids, titanium tweezers are preferable to prevent problems with charging of the grid.
3. Purchasing uranyl acetate can be expensive as some countries, such as Australia, have a high Dangerous Goods surcharge. An alternative is to use the protocol as above but with phosphotungstic acid. Make a 2% (w/v) aqueous solution of phosphotungstic acid and adjust the pH to approximately 7 using 0.5 M potassium hydroxide. This has to be done carefully as the phosphotungstic acid is not buffered.
4. Be aware that uranyl acetate and other heavy metal salts are cumulatively toxic. Uranyl acetate is also radioactive. Refer to your laboratory protocols for handling, storage and disposal of all radioactive materials and the appropriate Materials Safety and Data Sheet for heavy metal salts. Make sure that all pipettes are washed after use and filter paper, tips, etc. are disposed of appropriately.
5. Ethane is a highly flammable gas and should be kept away from any possible ignition sources.
6. While solutions containing low-volume fractions of solvents can be used, such as the conditions used to form $TTR_{105-115}$ fibrils (10% (v/v) acetonitrile and 90% (v/v) water), highly concentrated solvents can dissolve the film coating the grid,

and it is suggested that controls should be conducted with these solvents to determine their suitability for TEM analysis.

7. Some samples may benefit from the fixation in a solution of 1% EM-grade glutaraldehyde for 1 min. However, not all samples benefit and some may be distorted.
8. Glow discharging renders the hydrophobic surface of the carbon film hydrophilic and enables the sample and stain to adhere better to the carbon film ensuring that the sample spreads evenly. The effect of glow discharging does not last very long, depending on the ambient humidity up to half an hour, so process only a few grids at a time and prepare them immediately.
9. An alternative method for sample application, sample washing and the application of negative stain is as follows:
   (a) Lay down a strip of hydrophobic barrier film, such as parafilm (SPI supplies) and place 2×50 μl drops of water and 1×50 μl droplets of 2% (w/v) uranyl acetate on the surface.
   (b) Clamp a glow-discharged formvar/carbon-coated grid in a pair of fine tweezers.
   (c) Place the tweezers on a bench so that the carbon surface is facing up.
   (d) Apply 5 μl of sample to the carbon surface of the grid and allow the sample to adsorb for 1 min.
   (e) Drain the sample by picking up the grid and gently blotting with the grid edge-on to the filter paper. Do not allow the grid to completely dry.
   (f) Wash the grid by dipping it, face (sample side) down into a drop of water. Do not fully submerge the grid. Immediately remove, blot as before and move to the second drop, remove and blot.
   (g) After the second water wash, touch the grid to the surface of the stain drop and hold for 15 s. Remove and blot; this time, you should remove all the liquid and carefully blot between the tweezer tips to remove any liquid which might be lodged there.
   (h) Allow the grid to fully dry before inserting it into the microscope for viewing.
10. Uranyl acetate-stained grids can be kept in storage for at least a few weeks. Phosphotungstate is hydroscopic and unless grids are stored under vacuum, they should be looked at on the same day. Storage with desiccant may help.
11. You need to check with the scientist or engineer who maintains the microscope whether the scale bar has been accurately calibrated using a magnification standard. Commercial

cross-gratings can be applied for lower magnification and a suitable crystalline standard, such as catalase crystals or tobacco mosaic virus (TMV) with known lattice spacing, can be used for higher magnification. Calibration needs to be performed periodically to ensure accuracy.

12. A higher magnification image allows features, such as the fibril width, to be measured more accurately, although measurement of longer features, such as fibril length, may require images with lower magnification.
13. You will need to measure a reasonable number of samples to provide a good statistical representation if you are to comment on a characteristic of a population of fibrils, e.g. 100 fibrils or more, or to comment on the width distribution.
14. When using negative staining to enhance contrast, the edge of a fibril or other aggregate is defined by the presence of the stain along the edge of the structure. When the width of the fibril is measured from one side to another, the measurement also includes a width contribution from the stain. For this reason, the width or other measurements that rely on the stain to define the feature are an estimate with a small error associated with each measurement.
15. The concentration of the sample is critical for Cryo-EM; this can first be checked by negative stain and normal TEM (see Subheading 3.1). If the concentration is too high, the sample may tend to clump or aggregate, which can cause the sample to adhere to the filter paper during blotting, resulting in a low concentration or no sample on the grid.
16. Buffer salts or the presence of sucrose can interfere with the freezing process. Try to keep the salt concentration below 150 mM.
17. Once quick-frozen, the sample must then remain at liquid nitrogen temperature at all steps to the microscope.
18. With an FEI compustage, it is sometimes useful to pre-tilt the stage to −60° so that the holder opening is not completely inverted, thus retaining nitrogen in the Dewar during the pump sequence.

References

1. Goldsbury, C., Goldie, K., Pellaud, J., Seelig, J., Frey J., Muller, S.A., Kistler, J., Cooper, G.J.S., Aebi, U. (2000) Amyloid Fibril Formation from Full-Length and Fragments of Amylin. *Journal of Structural Biology.* **130**, 352–362.
2. Lashuel, H.A., Hartley, D.M., Petre, B.M., Wall, J.S., Simon, M.N., Walz, T., Lansbury Jr, P.T. (2003) Mixtures of Wild-type and a Pathogenic (E22G) Form of Aβ40 In Vitro Accumulate Protofibrils, Including Amyloid Pores. *Biochemistry.* **332**, 795–808.
3. Nieva, J., Shafton, A., Altobell III, L.J., Tripuraneni, S., Rogel, J.K., Wentworth, A.D., Lerner, R.A., Wentworth Jr., P.W. (2008) Lipid-derived aldehydes accelerate light chain amyloid and amorphous aggregation. *Biochemistry.* **47**, 7695–7705.

4. Sorci, M., Grassucci, R.A., Hahn, I., Frank, J., Belfort, G. (2009) Time-dependent insulin oligomer reaction pathway prior to fibril formation: Cooling and seeding. *Proteins.* **77,** 62–73.
5. Krysmann, M.J., Castelletto, V., Kelarakis, A., Hamley, I.W., Hule, R.A., Pochan, D.J. (2008) Self-Assembly and Hydrogelation of an Amyloid Peptide Fragment. *Biochemistry.* **47,** 4597–4605.
6. Engel, M.F.M., Khemte'mourian, L., Kleijer, C.C., Meeldijk, H.D., Jacobs, J., Verkleij, A.J., de Kruiff, B., Killan, J.A., Hoppener J.W.H. (2008) Membrane damage by human islet amyloid polypeptide through fibril growth at the membrane. *PNAS.* **105,** 6033–6038.
7. Thorn, D.C., Ecroyd, H., Sunde, M., Poon, S., Carver, J.A. (2008) Amyloid Fibril Formation by Bovine Milk α_{S2}-Casein Occurs under Physiological Conditions Yet Is Prevented by It's Natural Counterpart, α_{S1}-Casein. *Biochemistry.* **47,** 3926–3936.
8. Nilsson, M.R. (2004) Techniques to study amyloid fibril formation in vitro. *Methods.* **34,** 151–160.
9. Serpell, L.C., Sunde, M., Fraser, P.E., Luther, P.K., Morris, E.P., Sangren, O., Lundgren, E., Blake, C.C.F. (1995) Examination of the Structure of the Transthyretin Amyloid Fibril by Image Reconstruction from Electron Micrographs. *Journal of Molecular Biology.* **254,** 113–118.
10. Inoue, S., Kuroiwa, M., Saraiva, M.J., Guimaraes, A., Kisilevsky R. (1998) Ultrastructure of Familial Amyloid Polyneuropathy Amyloid Fibrils: Examination with High-Resolution Electron Microscopy. *Journal of Structural Biology.* **124,** 1–12.
11. Jimenez, J.L., Nettleton, E.J., Bouchard, M., Robinson, C.V., Dobson, C.M., Saibil, H.R. (2002) The protofilament structure of insulin amyloid fibrils. *Proceedings of the National Academy of Sciences.* **99,** 9196–9201.
12. Binger, K.J., Pham, C.L.L., Wilson, L.M., Bailey, M.F., Lawrence, L.J., Schuck P., Howlett, G.J. (2008) Apolipoprotein C-II Amyloid Fibrils Assemble via a Reversible Pathway that Includes Fibril Breaking and Rejoining. *Journal of Molecular Biology.* **376,** 1116–1129.
13. Gras, S.L., Tickler, A.K., Squires A.M., Devlin, G.L., Horton, M.A., Dobson, C.M., MacPhee, C.E. (2008) Functionalised amyloid fibrils for roles in cell adhesion. *Biomaterials.* **29,** 1553–62.
14. Goldsbury, C. Kistler, J., Aebi, U., Arvinte, T., Cooper, G.J.S. (1999) Watching Amyloid Fibrils Grow by Time Lapse Atomic Force Microscopy. *Journal of Molecular Biology.* **285,** 33–39.
15. Hayat, M.A. (ed) (2000) *Principles and Techniques of Electron Microscopy* (4th Ed) Cambridge University Press, Cambridge, UK.
16. Cohen-Krausz, S., Saibil, H.R. (2006) Three-dimensional Structural Analysis of Amyloid Fibrils by Electron Microscopy in *Protein Reviews* (Uversky, V.N., Fink A.L., ed.) Springer, USA, pp. 303–313.
17. Bremer, A., Henn, C., Engel, A., Baumeister, W., and Aebi, U., Has Negative Staining still a place in biomacromolecular electron microscopy? Ultramicroscopy 46, 85–111, 1992.
18. Taylor, K.A., Glaeser, R.M. (1974) Electron diffraction of frozen, hydrated protein crystals. *Science.* **186,** 1036–7.
19. Dubochet, J., McDowall, A.W. (1981) Vitrification of pure water for electron microscopy. *Journal of Microscopy.* **124,** RP3-RP4.
20. Adrian, M., Dubochet, J., Fuller, S.D., Harris, R.J. (1998) Cryo-negative staining. *Micron.* **29,** 145–160.
21. De Carlo, S., El-Bez, C., Alavarez-Rua, C., Borge, J., Dubochet, J. (2002) Cryo-negative staining reduces electron-beam sensitivity of vitrified biological particles. *Journal of Structural Biology.* **138,** 216–226.
22. Cavalier, A., Spehner, D., Humbel, B.M. (ed.) (2009) *Handbook of Cryo-Preparation Methods for Electron Microscopy,* CRC Press, Taylor and Francis Group, Boca Raton, FL, USA.

Chapter 14

Surface Plasmon Resonance Spectroscopy: A New Lead in Studying the Membrane Binding of Amyloidogenic Transthyretin

Xu Hou, David H. Small, and Marie-Isabel Aguilar

Abstract

Surface plasmon resonance (SPR) employs the optical principle of SPR to measure changes in mass on a sensor chip surface in real time. Surface chemistry has been developed which enables the immoblization of lipid bilayers and determination of protein–membrane interactions in real time. In the last decade, the plasma membrane has been demonstrated to play an important role in amyloidogenesis and cytotoxicity induced by amyloidogenic proteins. SPR provides an ideal way to study the membrane binding of amyloidogenic proteins. In this chapter, we describe the application of SPR to the study of amyloidogenic transthyretin binding to the plasma membrane and artificial lipid bilayers.

Key words: Amyloid, Transthyretin, Surface plasmon resonance, BIACORE biosensor, Plasma membrane, Lipid vesicle, Binding, Aqueous two-phase partition system

1. Introduction

Surface plasmon resonance (SPR) is a technique for real-time measurement of biomolecular interactions on biomimetic surfaces based on the optical phenomenon of SPR (1). SPR occurs under the condition of total internal reflection by thin layers of certain metals (e.g. gold). When a polarized laser light is directed through a medium with high refractive index (often a prism) to a thin layer of gold that lies on the border with a medium of low refractive index, a fraction of the light energy can penetrate the metal layer and excite electromagnetic waves (surface plasmons) that propagate within the metal surface at the interface with the sample solution. At a critical angle of incident light (the SPR

Andrew F. Hill et al. (eds.), *Protein Folding, Misfolding, and Disease: Methods and Protocols*, Methods in Molecular Biology, vol. 752, DOI 10.1007/978-1-60327-223-0_14, © Springer Science+Business Media, LLC 2011

angle), the SPR occurs in which energy of the incident light can be transferred to plasmons, resulting in a reduction in the intensity of reflected light (2). The SPR angle is dependent on the refractive index within a few hundred nanometres of the surface (1). Based on this principle, the SPR technique has been developed in which changes in the refractive index of the solution close to the sensor surface, which can be induced by binding of molecules onto the surface, are monitored in real time without the need for labelling (3). A typical SPR system consists of an SPR detector, light source, flow channel and sensor surface, comprising a conducting surface, such as gold or silver. Polarized light is emitted by the light source and reflected on the gold-coated sensor surface and detected by the diode array detector (Fig. 1a).

By automating the SPR measurement, commercial SPR spectroscopy instruments (e.g. the Biacore biosensors) can be used to analyze a large number of samples. In a Biacore biosensor, one of the interacting partners (i.e. the ligand) is immobilized onto the surface of a sensor chip that is coated by a thin layer of gold, while the sample solution containing the other partner (i.e. the analyte) is injected over the chip surface at a constant flow rate through a microfluidic channel system (4) (Fig. 1a) (refer to http://www.biacore.com). Changes in mass concentration at the chip surface arising from association and dissociation between the ligand molecules and the analyte molecules are measured as an SPR response unit (RU), and displayed in a sensorgram as a function of time (Fig. 1b). There is a linear relationship between the mass concentration at the chip surface and the SPR response unit such that 1 RU = 1 pg/mm^2 (5). The size of the change in the SPR response can thus be interpreted in terms of the stoichiometry of the interaction. In addition, analysis of the rate of change in the SPR response generates the apparent rate constants for the association and dissociation phases of the interaction, which can be used to calculate the apparent equilibrium constant, a measure of affinity. Owing to their high sensitivity, Biacore biosensors only require very small amounts of samples (nanomolar concentrations in some cases) for assays of molecular interactions.

Since its first application in characterizing antibody–antigen interactions in 1990 (6), Biacore biosensors have been used extensively in the research of molecular recognition, such as drug discovery and signalling transduction, and in genomic and proteomic studies (7–12). The application of SPR in the research of amyloidosis mainly focuses on two areas: the molecular mechanism of amyloidogenesis (refer to a recent review by Aguilar and Small (13)), and the interaction between membranes and amyloidogenic proteins. Since the invention of the L1 sensor chip which contains a carboxymethyldextran hydrogel derivatized with lipophilic alkyl chain anchors to capture lipid vesicles (Fig. 1a) (14), there has been a steady increase in the application of Biacore biosensors in studies on protein–membrane interactions (15–20).

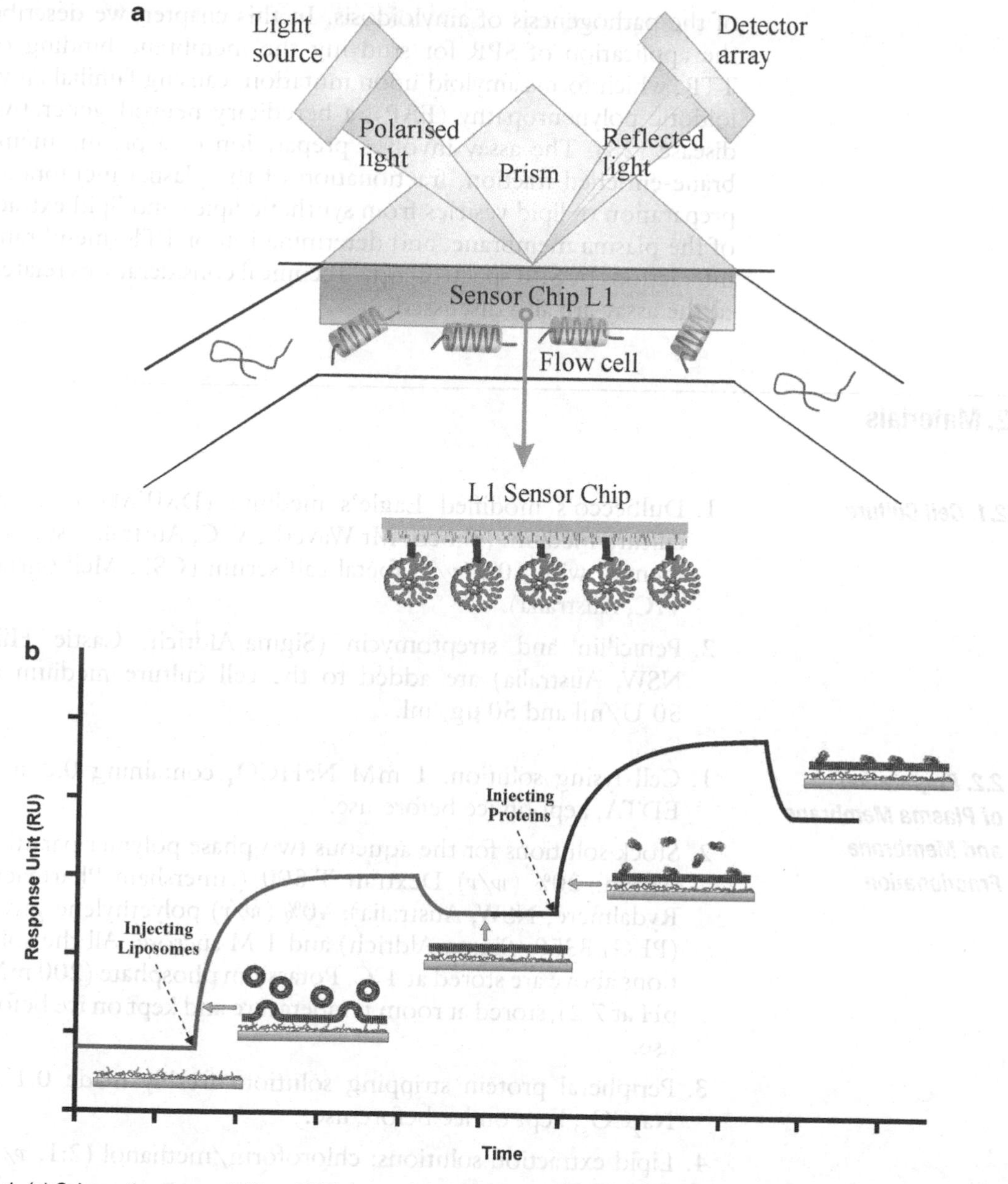

Fig. 1. (**a**) Schematic representation of SPR for protein–membrane interactions. The angle of incident light at which SPR occurs is dependent on the refractive index of the solution close to the surface of the sensor chip. Liposomes are immobilized onto an L1 sensor chip by lipophilic anchors attached to a dextran matrix on the chip surface. Protein samples are then injected over the sensor chip surface at a constant flow rate. Variations in mass concentration at the chip surface caused by protein binding are measured in real time by changes in the refractive index of the solution. (**b**) Diagram showing the changes in the SRP response (RU) as a function of time upon immobilizing liposomes on the surface of the sensor chip and binding of proteins on the mimetic membrane, including both the association and dissociation phases.

These biosensors have been used to examine the interactions between amyloidogenic proteins, such as amyloid β-protein (Aβ), prion protein (PrP) and transthyretin (TTR), and membranes (including the plasma membrane and artificial lipid membranes) (13, 21–27). The SPR has significantly expanded our understanding of the pathogenesis of amyloidosis. In this chapter, we describe the application of SPR for studying the membrane binding of TTR, which forms amyloid upon mutation, causing familial amyloidotic polyneuropathy (FAP), a hereditary neurodegenerative disease (28). The assay involves preparation of a plasma membrane-enriched fraction, fractionation of the plasma membrane, preparation of lipid vesicles from synthetic lipids and lipid extract of the plasma membrane, and determination of TTR–membrane interactions by SPR spectroscopy. Technical considerations related to the assay are also discussed.

2. Materials

2.1. Cell Culture

1. Dulbecco's modified Eagle's medium (DMEM)/F12 cell culture medium (Gibco, Mt Waverly, VIC, Australia) supplemented with 10% (*v/v*) foetal calf serum (CSL, Melbourne, VIC, Australia).
2. Penicillin and streptomycin (Sigma-Aldrich, Castle Hill, NSW, Australia) are added to the cell culture medium at 50 U/ml and 50 μg/ml.

2.2. Preparation of Plasma Membrane and Membrane Fractionation

1. Cell lysing solution: 1 mM $NaHCO_3$ containing 0.2 mM EDTA, kept on ice before use.
2. Stock solutions for the aqueous two-phase polymer partition system: 20% (*w/v*) Dextran T-500 (Amersham-Pharmacia, Rydalmere, NSW, Australia); 40% (*w/v*) polyethylene glycol (PEG) 3350 (Sigma-Aldrich) and 1 M sucrose. All the solutions above are stored at 4°C. Potassium phosphate (200 mM, pH at 7.2), stored at room temperature and kept on ice before use.
3. Peripheral protein stripping solution: freshly made 0.1 M Na_2CO_3, kept on ice before use.
4. Lipid extraction solutions: chloroform/methanol (2:1, *v/v*); 1 M H_2SO_4.

2.3. Preparation of Lipid Vesicles

1. Stock solutions of synthetic lipids: dimyristoyl-l-R-phosphatidylcholine (DMPC), dimyristoyl-l-R-phosphatidylserine (DMPS), dimyristoyl-l-R-phosphatidylethanolamine (DMPE), and sphingomyelin (SM) (Avanti Polar Lipids, Alabaster, AL) are dissolved in a chloroform/methanol mixture (1:1, *v/v*) at 2 mM. Cholesterol (CL) is dissolved in chloroform/methanol (3:1, *v/v*) at 2 mM. All lipid solutions are stored at −20°C.
2. Resuspension buffer: 20 mM sodium phosphate (pH 7.4) containing 150 mM NaCl.

3. Polycarbonate filter membranes with a pore size of 100 nm (Avestin, Ottawa, ON).

2.4. SPR Spectroscopy

1. Running buffer: 20 mM sodium phosphate (pH 7.4) containing 150 mM NaCl.
2. Cleansing buffer: 40 mM CHAPS; 2-propanol/50 mM NaOH (4:6, *v/v*).
3. Regeneration buffer: 10 mM NaOH.

3. Methods

3.1. Preparation of a Plasma Membrane-Enriched Fraction

A plasma membrane-enriched fraction of SH-SY5Y human neuroblastoma cells is prepared by means of an aqueous polymer-based two-phase partition system consisting of two aqueous polymers, dextran T-500 and PEG 3350 (29). The plasma membrane fraction is concentrated in the upper phase due to its higher affinity for PEG 3350. As determined by the activities of marker enzymes, there is a 3.6-fold enrichment in the plasma membrane, with approximately 80% mitochondria and 70% ER removed from the plasma membrane-enriched fraction (Table 1). The purity of the plasma membrane-enriched fraction is >85%. The purity of this fraction is comparable to that of a plasma membrane-enriched fraction of neurons isolated by density-gradient centrifugation (30, 31).

Experimental procedures (Fig. 2):

1. SH-SY5Y human neuroblastoma cells (approx. $6–8 \times 10^8$ cells) are harvested by centrifugation at $3,000 \times g$ for 5 min.

Table 1
Characterization of the plasma membrane fraction by marker enzymes

Marker enzyme	Cell homogenate ($n = 3$)	Plasma membrane fraction ($n = 3$)
p-Nitrophenylphosphatase[a] (μmol/h/mg protein)	0.21 ± 0.04	0.76 ± 0.11
Succinate cytochrome *c* reductase[b] (μmol/h/mg protein)	1.05 ± 0.05	0.19 ± 0.03
NADPH cytochrome *c* reductase[c] (nmol/h/mg protein)	9.96 ± 0.98	2.73 ± 0.61

[a]Marker enzyme for plasma membrane
[b]Marker enzyme for mitochondria
[c]Marker enzyme for endoplasmic reticulum

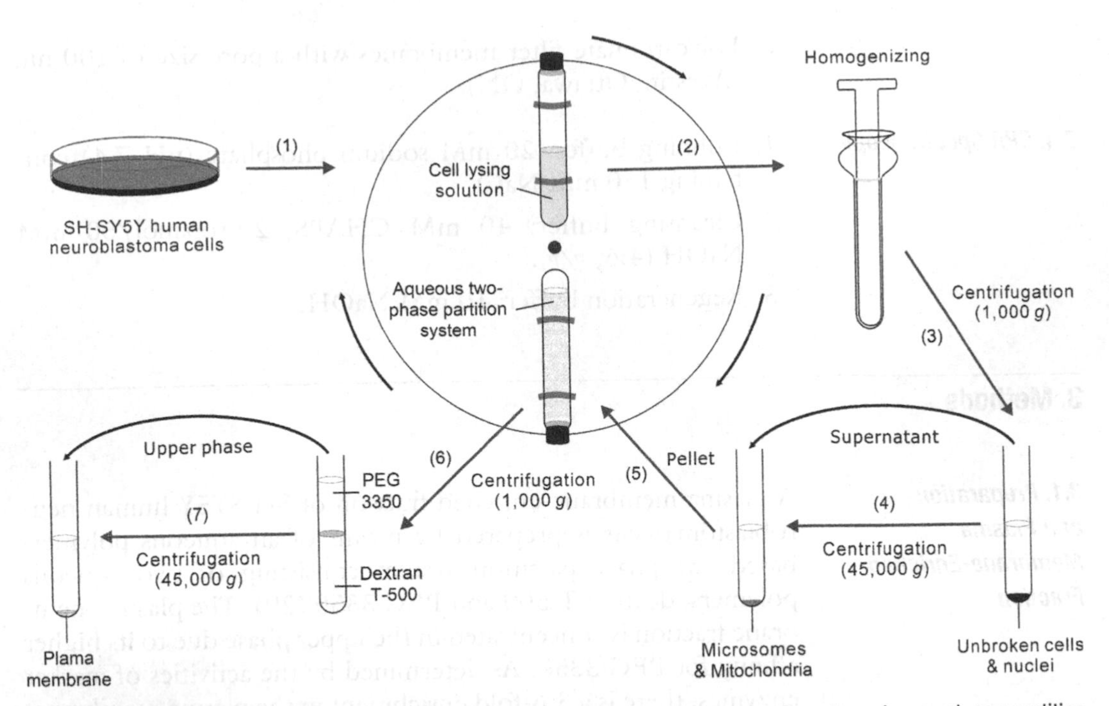

Fig. 2. Diagram of the purification protocol of the plasma membrane using an aqueous two-phase polymer partition technique. (1) SH-SY5Y human neuroblastoma cells collected by centrifugation are resuspended and mixed in the cell lysing solution to facilitate cytolysis; (2) the cells are then homogenized in a glass–glass homogenizer; (3) the cell lysate is centrifuged at 1,000 × *g* to remove unbroken cells and nuclei in the pellet; (4) the supernatant fraction is subject to ultracentrifugation at 45,000 × *g* to collect the microsome fraction; (5) the microsome fraction is then resuspended in an aqueous two-phase partition system containing dextran T-500 and PEG 3350, and mixed thoroughly; (6) the phases are separated by centrifugation at 1,000 × *g*; (7) finally, a plasma membrane-enriched fraction is extracted from the upper phase by ultracentrifugation at 45,000 × *g*.

The cell pellet is resuspended in the cell lysing buffer (1 mM $NaHCO_3$ containing 0.2 mM EDTA) at the density of 10^8/ml. The cell suspension is then gently shaken on a rotator (at ~20 rpm) for 40 min at 4°C (see Note 1).

2. The cells are disrupted with a Dounce glass–glass homogenizer (Kontes, Vineland, NJ) for ~70 strokes on ice. Cell lysis is examined by staining with 0.4% trypan blue (see Note 2).
3. The cell lysate is centrifuged at 1,000 × *g* at 4°C for 2 × 10 min to remove unbroken cells and nuclei (in the pellet).
4. The supernatant is further centrifuged at 45,000 × *g* for 30 min at 4°C to collect the microsomal fraction.
5. The pellet (containing microsomes and mitochondria) is resuspended in 50 mM Tris–HCl (pH 8.0) to 8 g. An aliquot of cell suspension (2 g) is added to the aqueous two-phase partition system (16 g in total, composition shown in Table 2) consisting of 6.6% (*w/w*) dextran T-500, 6.6% (*w/w*) PEG 3350, 250 mM sucrose, and 5 mM potassium phosphate (pH 7.2). The two-phase partition system is then mixed on a rotator at 4°C for 40 min.

Table 2
Composition of the aqueous two-phase partition system

Component	Amount added	Final concentration
20% Dextran T-500	5.12 g	6.6% (*w/w*)
40% PEG 3350	2.56 g	6.6% (*w/w*)
200 mM Potassium phosphate (pH 7.2)	0.4 ml	5 mM
1 M Sucrose	1.6 ml	250 mM
Microsomal fraction	2 g	–

Add Milli-Q H_2O to a total weight of 16 g

6. Phase separation is achieved by centrifugation at 1,000 × *g* for 10 min at 4°C.
7. The upper phase containing PEG 3350 is transferred to another centrifuge tube (see Note 3) and diluted by fivefold in 50 mM Tris–HCl (pH 8.0). The plasma membrane fraction is collected by centrifugation at 45,000 × *g* for 30 min at 4°C.
8. The pellet is finally resuspended in 20 mM sodium phosphate (pH 7.4) containing 150 mM NaCl at a protein concentration of 0.2–0.3 mg/ml (as determined by the BCA assay). The plasma membrane preparation is stored at −80°C.

3.2. Fractionation of the Plasma Membrane Preparation

The peripheral membrane-associated proteins are removed by sodium carbonate treatment, adapted from the protocol described in Fujiki et al. (32).

3.2.1. Removal of Peripheral Membrane Proteins

Experimental procedures:

1. An aliquot (500 μl) of the plasma membrane preparation in 50 mM Tris–HCl (pH 8.0) is thawed at 37°C in a water bath and then centrifuged at 45,000 × *g* for 30 min at 4°C.
2. The pellet is resuspended in 200 μl of freshly made, prechilled 0.1 M Na_2CO_3 and kept on ice for 15 min.
3. The membrane preparation is centrifuged at 52,000 × *g* for 10 min at 4°C. The supernatant, which contains peripheral membrane proteins, is removed.
4. The pellet is again resuspended in 200 μl of 0.1 M Na_2CO_3. The steps 2 and 3 are repeated a second time.
5. The resulting plasma membrane stripped of peripheral proteins is resuspended in 20 mM sodium phosphate (pH 7.4) containing 150 mM NaCl at a protein concentration of 0.2–0.3 mg/ml and stored at −80°C.

3.2.2. Extraction of Membrane Lipids

Membrane lipids are extracted using organic solvents based on the protocol described in Folch et al. (33).

Experimental procedures:

1. An aliquot (500 μl) of the plasma membrane preparation is thawed at 37°C in a water bath and then centrifuged at 45,000 × *g* for 30 min at 4°C.
2. The pellet is resuspended in 8 ml chloroform/methanol (2:1, *v/v*) and mixed thoroughly on a rotator (~30 rpm) at 4°C overnight.
3. H_2SO_4 (1 ml of 1 M solution) is added to the solution followed by thorough vortexing. The solution is then centrifuged at 500 × *g* for 10 min at 4°C.
4. The lower phase (chloroform containing lipid extract) is transferred to another tube (see Note 4).
5. H_2SO_4 (1 ml of 1 M solution) is added to the remaining upper phase. The steps 3 and 4 are repeated a second time.
6. Aliquots (1 ml each) of the lipid extract are dried under a stream of N_2 gas and then further dehydrated in vacuo overnight (see Note 5). The dehydrated lipid extract is stored at −20°C.

3.3. Preparation of Lipid Vesicles

3.3.1. Preparation of Synthetic Lipids

1. Stock solutions (2 mM) of synthetic DMPC, DMPE, DMPS, SM, and CL are mixed at a ratio of 6:5:2:2:5 to mimic the lipid composition of human neural tissues (34).
2. Aliquots (408 μl) of the lipid mixture are dried under a stream of N_2 gas and then further dehydrated in vacuo overnight. The dehydrated lipid samples are stored at −20°C.

3.3.2. Preparation of Lipid Vesicles

1. An aqueous solution, 600 μl of 20 mM sodium phosphate (pH 7.4) containing 150 mM NaCl, is added to the dried lipids (either synthetic or from the plasma membrane).
2. The lipid suspension is sonicated in a water bath sonicator until the solution becomes clear (see Note 6).
3. Unilamellar lipid vesicles are prepared by extrusion through an Avestin polycarbonate filter membrane with a 100 nm pore size using a LiposoFast extruder (Avestin, Ottawa, ON) for 17 times (see Note 7).

3.4. Determination of TTR–Membrane Interactions by SPR Spectroscopy

1. After an L1 sensor chip is docked into a Biacore 3000 biosensor (Biacore AB, Uppsala, Sweden), the chip is conditioned at room temperature by 10 μl of 40 mM CHAPS at 50 μl/min to clean the surface. The prime procedure, which flushes the instrument fluidic system and chip surface, is performed twice.

2. Lipid vesicles (100 μl) in 20 mM sodium phosphate (pH 7.4) containing 150 mM NaCl or plasma membrane preparations (100 μl) are immobilized onto the sensor chip surface by passing over the surface at a flow rate of 2 μl/min at room temperature. The plasma membrane preparation and carbonate-treated plasma membrane are immobilized onto the sensor chip surface in a similar manner (see Note 8).
3. The running buffer, 20 mM sodium phosphate (pH 7.2) containing 150 mM NaCl, is passed over the surface for 10 min after the end of injection to allow for equilibration.
4. NaOH (10 μl of 10 mM solution) is applied at a rate of 50 μl/min to remove multilamellar lipids on the sensor chip surface. Successful coating of the chip surface is demonstrated by an increase in the instrument RU (Fig. 3a) (see Note 9).
5. A TTR solution (100 μl, either freshly made or aged) in 20 mM sodium phosphate (pH 7.4) containing 150 mM NaCl (see Note 10) is applied to the immobilized membrane at 30 μl/min over a period of 200 s. TTR is replaced by the running buffer at the end of injection for a further 10 min so that the dissociation of TTR can be monitored. The amount of binding is measured from the change in RU in the sensorgrams (Fig. 3b).
6. To remove residual TTR from the membrane surface after dissociation for 10 min, NaOH (10 μl of a 10 mM solution) is injected at 50 μl/min (see Note 11).
7. A series of TTR solutions ranging from 0.1 to 40 μM are applied to the immobilized membrane. The binding kinetics are determined by fitting the binding curves to different predefined numerical models (Fig. 4) (see Note 12).
8. At the end of the assay, 10 μl of 40 mM CHAPS and 2-propanol/50 mM NaOH (4:6, v/v) are injected at 50 μl/min to strip the membrane bilayer off the sensor chip (see Note 13).

4. Notes

1. Suspension in the hypotonic cell lysing solution results in cell swelling, facilitating cell disruption.
2. After homogenizing for ~70 strokes, at least 80% of the cells are expected to disrupt as shown by trypan blue staining. If cell disruption is not satisfactory, another ~30 strokes can be performed until 80% or more of the cells are lysed. It is critical to keep the cell lysate on ice all the time.
3. It is critical not to disturb the interface and lower phase.

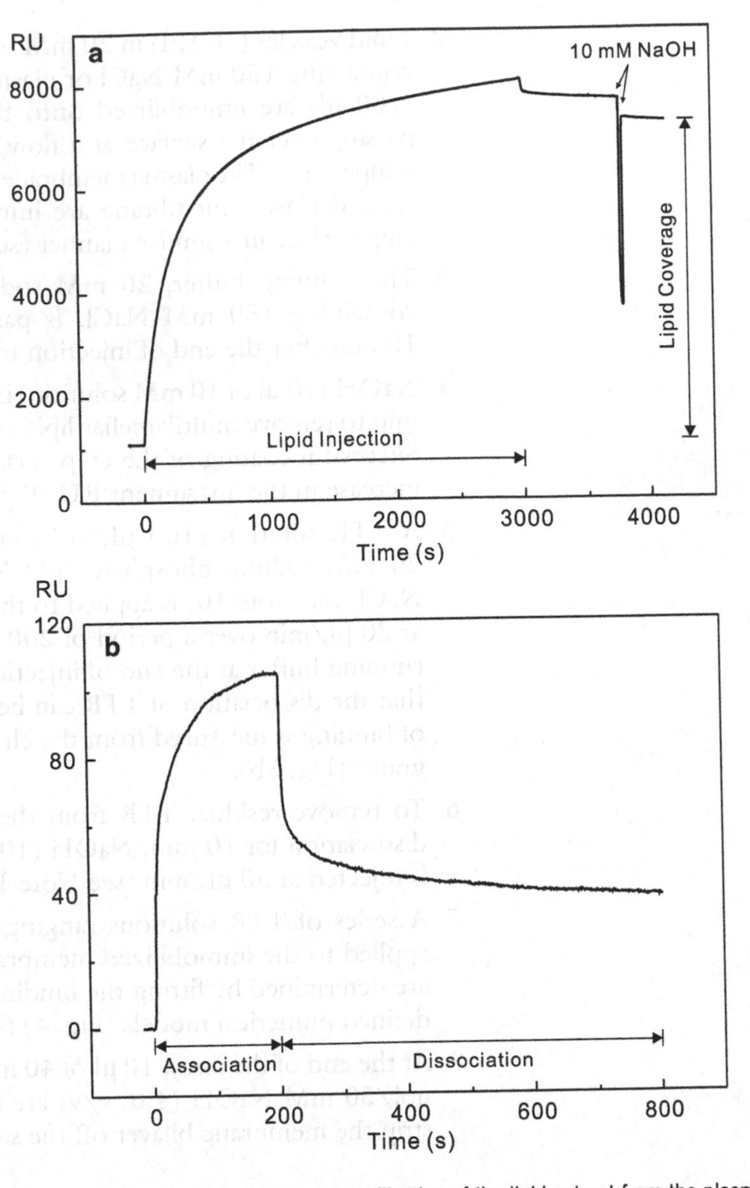

Fig. 3. (**a**) Sensorgram showing the immobilization of the lipid extract from the plasma membrane onto the surface of an L1 sensor chip. (**b**) Sensorgram showing the binding of a freshly made solution (10 μM) of an amyloidogenic TTR mutant, L55P TTR, to the immobilized lipid extract from the plasma membrane.

4. The transfer pipette is slowly pressed to the bottom of the tube before sucking the solution. It is critical not to disturb the upper phase and interface. Wipe the outer side of the transfer pipette before injecting to another tube.
5. By drying the lipids, a uniform lipid film adhering to the glass that can form lipid vesicles upon hydration is created.

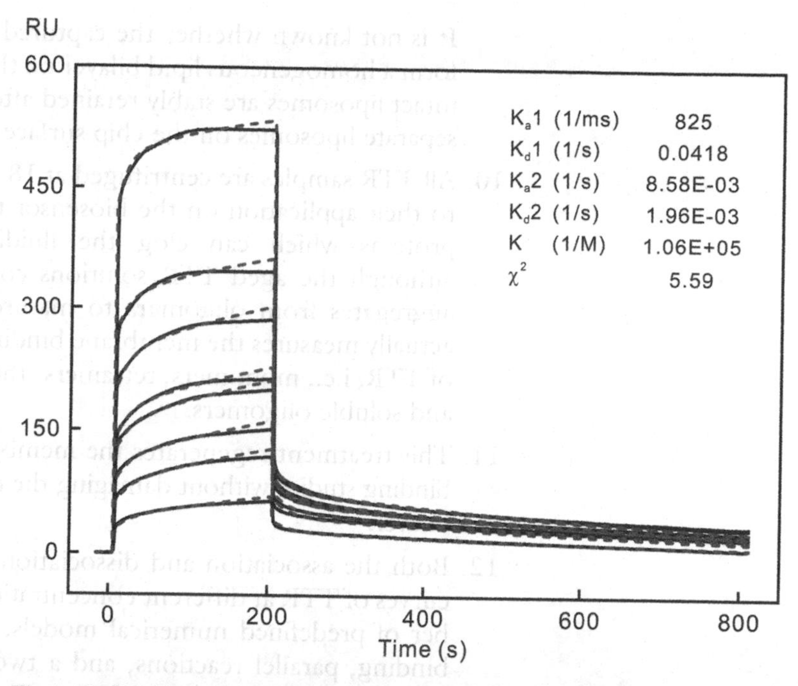

Fig. 4. Binding kinetics are determined by fitting the sensorgrams produced from a series of analyte concentrations with various numerical models using the BIAevaluation software. Shown here are sensorgrams (*solid lines*) of freshly made L55P TTR (1–40 μM) binding to lipid extract of the plasma membrane. *Dashed lines* represent the best fitted numerical binding model, the two-state reaction model, at these concentrations of TTR. The kinetic parameters calculated from the two-state reaction model are shown in the *upper right* (see Note 14).

6. To facilitate hydration of the lipids, it is necessary to let the water bath sonicator heat up gradually to beyond the transition temperature of the lipids (usually 40–50°C produces good hydration).
7. Multilamellar lipid micelles can be disrupted by the extrusion procedure. Empirically, 17 times of extrusion produces a satisfactory uniformity of the resulting lipid vesicles. It is necessary to extrude lipids at a temperature higher than the transition temperature. Therefore, if the resistance is too high, extrude in a water bath at 40–50°C.
8. The plasma membrane preparation and carbonate-treated plasma membrane are extruded through polycarbonate membranes with a pore size of 100 nm at 37°C prior to immobilization. They should not be sonicated as sonication may disrupt protein structures in the plasma membrane.
9. After the chip has been coated with lipids, an increase of 5,000–7,000 RU is expected, indicating complete coverage of the chip by the lipid bilayer (14, 35). In case the coverage is not satisfactory, a second injection of lipid vesicles is needed.

It is not known whether the captured lipid vesicles fuse and form a homogeneous lipid bilayer on the chip surface (35), or intact liposomes are stably retained after injection and stay as separate liposomes on the chip surface (14, 36).

10. All TTR samples are centrifuged at 18,000 × *g* for 1 min prior to their application on the biosensor to remove precipitated proteins which can clog the fluidic system. Therefore, although the aged TTR solutions contain a wide range of aggregates from oligomers to mature amyloid fibrils, SPR actually measures the membrane binding of the soluble forms of TTR, i.e., monomers, tetramers (the native form of TTR), and soluble oligomers.
11. This treatment regenerates the membrane bilayer for further binding studies without damaging the immobilized membrane bilayer.
12. Both the association and dissociation phases of the binding curves of TTR at different concentrations are fitted to a number of predefined numerical models, such as 1:1 Langmuir binding, parallel reactions, and a two-state reaction model, using the BIAevaluation software. Detailed interpretation of these numerical binding models can be found elsewhere (18, 37, 38). Binding kinetics are calculated from the model providing the best fit (lowest chi² (χ^2)).
13. After the cleansing procedure, the baseline is expected to return to the level before the binding assay. Other cleansing reagents can be used in combination with those mentioned in Subheading 3, such as 6 M guanidine–HCl, 0.1 M acetic acid and 0.2% SDS.
14. A two-state reaction model suggested that there is likely to be at least two steps involved in the interaction between TTR and the membrane systems. These steps may involve initial binding to the lipid head groups and then either further insertion into the hydrocarbon region of the membrane or a conformational change in TTR upon binding to the membrane (37, 38), consistent with previous reports on the membrane binding of Aβ (39, 40).

References

1. Markey, F. (2000) Principles of surface plasmon resonance, in *Real-time analysis of biomolecular interactions: applications of BIACORE, Ed.* (K. Nagata and H. Handa, ed.) Springer-Verlag, Tokyo, pp. 13–22.
2. Kretschmann, E. (1971) Die Bestimmung optischer Konstanten von metallen durch Anregung von Oberflachenplasmaschwingungen. *Z Phys.* **241**, 313–324.
3. Jonsson, U., Fagerstam, L., Ivarsson, B., Johnsson, B., Karlsson, R., Lundh, K., Lofas, S., Persson, B., Roos, H., Ronnberg, I. and et al. (1991) Real-time biospecific interaction analysis using surface plasmon resonance and a sensor chip technology. *Biotechniques.* **11**, 620–627.
4. Hashimoto, S. (2000) Principles of BIACORE, in *Real-time analysis of biomolecular interactions: applications of BIACORE, Ed.* (K. Nagata and H. Handa, ed.) Springer-Verlag, Tokyo, pp. 23–32.
5. Stenberg, E., Persson, B., Roos, H. and Urbaniczky, C. (1990) Quantitative

determination of surface concentration of protein with surface plasmon resonance by using radiolabeled proteins. *J Colloid Interface Sci.* **143**, 513–526.

6. Fagerstam, L. G., Frostell, A., Karlsson, R., Kullman, M., Larsson, A., Malmqvist, M. and Butt, H. (1990) Detection of antigen-antibody interactions by surface plasmon resonance. Application to epitope mapping. *J Mol Recognit.* **3**, 208–214.
7. Rich, R. L. and Myszka, D. G. (2000) Survey of the 1999 surface plasmon resonance biosensor literature. *J Mol Recognit.* **13**, 388–407.
8. Rich, R. L. and Myszka, D. G. (2001) Survey of the year 2000 commercial optical biosensor literature. *J Mol Recognit.* **14**, 273–294.
9. Rich, R. L. and Myszka, D. G. (2002) Survey of the year 2001 commercial optical biosensor literature. *J Mol Recognit.* **15**, 352–376.
10. Rich, R. L. and Myszka, D. G. (2003) Survey of the year 2002 commercial optical biosensor literature. *J Mol Recognit.* **16**, 351–382.
11. Rich, R. L. and Myszka, D. G. (2005) Survey of the year 2003 commercial optical biosensor literature. *J Mol Recognit.* **18**, 1–39.
12. Myszka, D. G. (1999) Survey of the 1998 optical biosensor literature. *J Mol Recognit.* **12**, 390–408.
13. Aguilar, M. I. and Small, D. H. (2005) Surface plasmon resonance for the analysis of beta-amyloid interactions and fibril formation in Alzheimer's disease research. *Neurotox Res.* **7**, 17–27.
14. Cooper, M. A., Hansson, A., Lofas, S. and Williams, D. H. (2000) A vesicle capture sensor chip for kinetic analysis of interactions with membrane-bound receptors. *Anal Biochem.* **277**, 196–205.
15. Plant, A. L., Brigham-Burke, M., Petrella, E. C. and O'Shannessy, D. J. (1995) Phospholipid/alkanethiol bilayers for cell-surface receptor studies by surface plasmon resonance. *Anal Biochem.* **226**, 342–348.
16. Besenicar, M., Macek, P., Lakey, J. H. and Anderluh, G. (2006) Surface plasmon resonance in protein-membrane interactions. *Chem Phys Lipids.* **141**, 169–178.
17. McDonnell, J. M. (2001) Surface plasmon resonance: towards an understanding of the mechanisms of biological molecular recognition. *Curr Opin Chem Biol.* **5**, 572–577.
18. Mozsolits, H. and Aguilar, M. I. (2002) Surface plasmon resonance spectroscopy: an emerging tool for the study of peptide-membrane interactions. *Biopolymers.* **66**, 3–18.
19. Cooper, M. A. (2004) Advances in membrane receptor screening and analysis. *J Mol Recognit.* **17**, 286–315.
20. Mozsolits, H., Thomas, W. G. and Aguilar, M. I. (2003) Surface plasmon resonance spectroscopy in the study of membrane-mediated cell signalling. *J Pept Sci.* **9**, 77–89.
21. Ariga, T., Kobayashi, K., Hasegawa, A., Kiso, M., Ishida, H. and Miyatake, T. (2001) Characterization of high-affinity binding between gangliosides and amyloid beta-protein. *Arch Biochem Biophys.* **388**, 225–230.
22. Inaba, S., Okada, T., Konakahara, T. and Kodaka, M. (2005) Specific binding of amyloid-β-protein to IMR-32 neuroblastoma cell. *J Pept Res.* **65**, 485–490.
23. Subasinghe, S., Unabia, S., Barrow, C. J., Mok, S. S., Aguilar, M. I. and Small, D. H. (2003) Cholesterol is necessary both for the toxic effect of Abeta peptides on vascular smooth muscle cells and for Abeta binding to vascular smooth muscle cell membranes. *J Neurochem.* **84**, 471–479.
24. Kremer, J. J. and Murphy, R. M. (2003) Kinetics of adsorption of beta-amyloid peptide Abeta(1–40) to lipid bilayers. *J Biochem Biophys Methods.* **57**, 159–169.
25. Critchley, P., Kazlauskaite, J., Eason, R. and Pinheiro, T. J. (2004) Binding of prion proteins to lipid membranes. *Biochem Biophys Res Commun.* **313**, 559–567.
26. Hou, X., Richardson, S. J., Aguilar, M. I. and Small, D. H. (2005) Binding of amyloidogenic transthyretin to the plasma membrane alters membrane fluidity and induces neurotoxicity. *Biochemistry.* **44**, 11618–11627.
27. Hou, X., Mechler, A., Martin, L. L., Aguilar, M. I. and Small, D. H. (2008) Cholesterol and anionic phospholipids increase the binding of amyloidogenic transthyretin to lipid membranes. *Biochim Biophys Acta.* **1778**, 198–205.
28. Hou, X., Aguilar, M. I. and Small, D. H. (2007) Transthyretin and familial amyloidotic polyneuropathy: clues to the molecular mechanism of neurodegeneration in amyloidosis. *FEBS J.* **274**, 1637–1650.
29. Morre, D. J. and Morre, D. M. (1989) Preparation of mammalian plasma membranes by aqueous two-phase partition. *Biotechniques.* 7, 946–948, 950–944, 956–948.
30. Henn, F. A., Hansson, H.-A. and Hamberger, A. (1972) Preparation of plasma membrane from isolated neurons. *J Cell Biol.* **53**, 654–661.
31. Mersel, M., Lelong, I., Hindelang, C., Sarlieve, L. L. and Vincendon, G. (1987) Isolation of plasma membranes from neurons grown in primary culture. *Anal Biochem.* **166**, 246–252.
32. Fujiki, Y., Hunnard, A. L., Fowler, S. and Lazarow, P. B. (1982) Isolation of intracellular membranes by means of sodium carbonate

treatment: application to endoplasmic reticulum. *J Cell Biol.* **93**, 97–102.
33. Folch, J., Lees, M. and Stanley, G. H. (1957) A simple method for the isolation and purification of total lipides from animal tissues. *J Biol Chem.* **226**, 497–509.
34. Suzuki, K. (1976) Chemistry and metabolism of brain lipids, in *Basic neurochemistry, 2nd Ed.* (G. J. Siegel, R. W. Albers, R. Katzman and B. W. Agranoff, ed.) Little, Brown and Company, Boston, pp. 308–328.
35. Erb, E. M., Chen, X., Allen, S., Roberts, C. J., Tendler, S. J., Davies, M. C. and Forsen, S. (2000) Characterization of the surfaces generated by liposome binding to the modified dextran matrix of a surface plasmon resonance sensor chip. *Anal Biochem.* **280**, 29–35.
36. Anderluh, G., Besenicar, M., Kladnik, A., Lakey, J. H. and Macek, P. (2005) Properties of nonfused liposomes immobilized on an L1 Biacore chip and their permeabilization by a eukaryotic pore-forming toxin. *Anal Biochem.* **344**, 43–52.
37. Henriques, S. T., Pattenden, L. K., Aguilar, M. I. and Castanho, M. A. (2008) PrP(106–126) does not interact with membranes under physiological conditions. *Biophys J.* **Epub**, doi:10.1529/biophysj.1108.131458.
38. Mozsolits, H., Wirth, H. J., Werkmeister, J. and Aguilar, M. I. (2001) Analysis of antimicrobial peptide interactions with hybrid bilayer membrane systems using surface plasmon resonance. *Biochim Biophys Acta.* **1512**, 64–76.
39. Kakio, A., Nishimoto, S., Yanagisawa, K., Kozutsumi, Y. and Matsuzaki, K. (2002) Interactions of amyloid beta-protein with various gangliosides in raft-like membranes: importance of GM1 ganglioside-bound form as an endogenous seed for Alzheimer amyloid. *Biochemistry.* **41**, 7385–7390.
40. Matsuzaki, K. and Horikiri, C. (1999) Interactions of amyloid beta-peptide (1–40) with ganglioside-containing membranes. *Biochemistry.* **38**, 4137–4142.

Chapter 15

Elucidating the Role of Metals in Alzheimer's Disease Through the Use of Surface-Enhanced Laser Desorption/Ionisation Time-of-Flight Mass Spectrometry

Andrew D. Watt, Keyla A. Perez, and Lin Wai Hung

Abstract

Alzheimer's disease (AD) is a highly heterogeneous and progressive dementia which is characterised by a progressive decline in cognitive functioning, selective neuronal atrophy, and loss of cortical volume in areas involved in learning and memory. However, recent research has indicated that the AD-affected brain is also besieged by increases in oxidative stress as well as perturbations to the homeostasis of biometals, such as copper and iron. These metals are known to interact with the neuropathological hallmark of AD, the β-amyloid peptide (Aβ), in a manner which increases Aβ's neurotoxic effects. This knowledge has led to the development of therapeutic measures which act to restore biometal homeostasis within the AD brain.

This chapter outlines how Surface-Enhanced Laser Desorption/Ionisation Time-of-Flight Mass Spectrometry can be used to monitor Aβ levels within biological systems as well as describing the use of immobilised metal affinity capture in the observation of synthetic Aβ peptides.

Key words: Alzheimer's disease, Aβ, Metal homeostasis, SELDI-TOF MS, Antibody capture, Immobilised metal affinity capture

1. Introduction

1.1. Alzheimer's Disease and Aβ

Alzheimer's disease (AD) is a highly heterogeneous and progressive neurodegenerative disorder typified by neuronal atrophy, loss of cortical volume, and decreased synaptic plasticity. On a neuropathological level, the AD-affected brain is characterised by the presence of extracellular amyloid plaques within the parahippocampal, inferior temporo-parietal, fusiform and orbito-frontal regions of the brain (1–3). These plaques are composed of aggregates of the amyloid β-polypeptide (Aβ) (4, 5), a 4.5 kDa peptide

Andrew F. Hill et al. (eds.), *Protein Folding, Misfolding, and Disease: Methods and Protocols*, Methods in Molecular Biology, vol. 752, DOI 10.1007/978-1-60327-223-0_15, © Springer Science+Business Media, LLC 2011

derived from the sequential cleavage of the type 1 membrane-spanning glycoprotein, aptly named the amyloid precursor protein (APP) (6–9). Proteolytic cleavage of APP is governed by α-, β-, and γ-secretases to produce two distinct pathways. The non-amyloidogenic pathway involves cleavage of APP by α-secretase within the Aβ sequence itself, thus preventing the formation of Aβ (10, 11). In the amyloidogenic pathway, however, APP is sequentially cleaved by β- and γ-secretases to produce Aβ peptides of between 39 and 43 amino acids in size (7). The most abundant derivative of APP cleavage is $A\beta_{1\text{-}40}$. However, it is the less abundant $A\beta_{1\text{-}42}$, which aggregates more readily into neuritic plaques and is thus considered to be more inherently toxic (12, 13).

The central role played by Aβ in the development and progression of AD has lead to the development of the amyloid cascade theory, the predominant etiological theory of AD pathogenesis (14). This theory purports that AD pathogenesis arises due to an inherent imbalance in the production and elimination of Aβ within the AD brain. This imbalance results in the accumulation of Aβ which is thought to initiate a series of downstream events, including: Aβ oligomerisation and subsequent aggregation into neuritic plaques, as well as the phosphorylisation of tau neurofibrillary tangles. These events culminate in the aberration of synaptic functioning, neuronal atrophy, and the onset of impairments in cognitive functioning (14, 15).

While the amyloid cascade theory readily accounts for the observable characteristics of AD, little remains known of the underlying cellular mechanisms which regulate the interwoven pathways of the cascade itself. In recent years, however, research has increasingly suggested that many of these pathways may be regulated through interactions between biologically available metals, such as copper and iron, and the by-products of the amyloid cascade.

1.2. The Role of Metals in Alzheimer's Disease

Biologically available transition metals are essential for normal cellular functions as their redox potential can be harnessed by enzymes to drive an array of metabolic processes (16). In the normal brain, the homeostasis of these metals is rigorously controlled, with effectively no passive flux of metal ions occurring across the blood brain barrier (17). However, as the brain ages, there is a gradual breakdown in the stringent regulatory mechanisms of metal homeostasis resulting in age-related increases in Cu and Fe levels in the brain (18–20). In the absence of these strict regulatory mechanisms, the accumulating transition metals are thought to actively damage neuronal cells through oxidative stress, a process culminating in the generation of the reactive oxygen species (ROS), such as the hydroxyl radical ($OH^{\bullet}$), via the Fenton and Haber–Weiss reactions (21).

The two most profuse redox-active metals are Cu and Fe (22, 23) and both have been shown to facilitate the production of hydrogen peroxide through their interactions with Aβ. The resulting hydrogen peroxide then acts to oxidise Aβ and produces cross-linked and soluble Aβ which is protease resistant (24, 25). Thus, it is not surprising that the roles of transition metals, such as Cu and Fe, are becoming increasingly more prominent in AD pathology.

1.3. Metal-Based Therapeutics for Alzheimer's Disease

The involvement of metals in the progression of AD has led to the development of therapeutic interventions centring on the restoration of metal homeostasis in the AD brain (26). One such potential therapeutic agent is Cu^{II}(gtsm), which acts to inhibit the activity of glycogen synthase kinase 3β (GSK3β) by increasing intracellular copper bioavailability (27). GSK3β was selected as a target in response to findings that GSK3β was capable of phosphorylating tau (28) and that inhibition of GSK3β resulted in an increase in Aβ degrading protease expression (29). In a study by Crouch et al. (27), Surface-Enhanced Laser Desorption/Ionisation Time-of-flight mass spectrometry (SELDI-TOF MS) was utilised to demonstrate that this inhibition of GSK3β by Cu^{II}(gtsm) resulted in significant decreases in oligomeric Aβ within the brains of transgenic APP/PS1 mice and subsequent restorations in cognitive performances.

The findings of this study act to demonstrate the intricate relationship between metal homeostasis and the underlying pathogenic processes of AD and provide additional support for the important role that metals are playing in our understanding of AD. However, they also demonstrate the efficacy of SELDI-TOF MS as a tool for further elucidating the role metals play within the AD-affected system.

1.4. Surface-Enhanced Laser Desorption/Ionisation Time-of-Flight Mass Spectrometry

The principles of SELDI-TOF MS are reasonably simple: proteins, or peptides, are captured atop a solid-phase ProteinChip® surface by way of adsorption, electrostatic interaction, partition, or affinity chromatography before they are co-crystallised on the chip surface with a matrix (30). The matrix is an energy absorbing molecule (EAM), such as α-cyano-4-hydroxycinnamic acid (CHCA), or sinapinic acid (SPA), which is used to assist in the ionisation and desorption of the proteins from the chips surface (31). Subsequent to the addition of the EAM, the sample is irradiated by the pulsed, UV, nitrogen laser resulting in sample desorption/ionisation. These species are then accelerated through the vacuum-sealed TOF tube into the ion detector. With this process, a mass to charge ratio (M/Z) for each protein/peptide in the sample is recorded on the basis of the TOF required for each species to pass through the TOF tube and reach the detector (30). SELDI-TOF MS thus separates proteins within the sample based on their chromatographic properties (32) and the differentially expressed proteins can be identified through the comparison of peak intensities in the spectra (31). The properties of the protein species caught are dependent upon the chromatographic interactions between the proteins in the sample and the specific chip surface utilised (33). The surfaces of the arrays are derivatised with standard chromatographic separation moieties, including ion exchange, reverse phase, and immobilised metal affinity capture (IMAC) (33).

In contrast to classic mass spectrometry techniques, the enhanced surfaces of the ProteinChip® arrays enable a more pronounced enrichment of the captured proteins through the heightened specificity of the molecular interactions occurring between the chip surface and the sample (33). This means that sample volumes as low as 0.5 μL can be analysed (34) and peaks containing as little as 50 fmol can be detected (35), which is essential for the detection of low abundance proteins in dynamic biological systems (36). Thus, when coupled with the broad binding properties of the IMAC ProteinChip® surface, SELDI-TOF MS provides a highly adaptive technique by which to undertake proteomic profiling (33).

1.4.1. Immobilised Metal Affinity Capture

IMAC operates through the reversible binding of proteins to the array surface by utilising coordinated metal interactions. The chip is primed with a metal, such as Cu or Fe, which forms a stable octahedral complex with the nitrilotriacetic acid (NTA) on the chip surface. This stable complex leaves two free sites with which a metallo-protein interaction may occur between the metal and specific amino acid residues on the proteins in the sample (37). Recently IMAC, in conjunction with SELDI-TOF MS, has been used to identify potential metallo-protein biomarkers for AD (38, 39). While further research is required to ascertain the importance of these biomarkers in the pathogenesis, and potential diagnosis of AD, their discoveries lend support to the use of both SELDI-TOF MS and metallo-protein interactions as means of creating a proteomic profile which distinguishes AD from healthy controls (HC).

The use of SELDI-TOF MS as a means of observing synthetic Aβ peptides through the use of antibody capture techniques has been previously described in the literature (40).

However, in this chapter, we expand on this knowledge by describing the methodology of observing Aβ peptides within biological systems. The high specificity of antibody capture is useful as an adjunct to therapeutic trials, where levels of a particular antigen are being monitored. In addition to this, the current chapter also describe techniques by which IMAC in conjunction with SELDI-TOF MS can be used to capture synthetic Aβ peptides. The IMAC methodology which follows can be thought of as a foundation for proteomic explorations within biological systems.

2. Materials

2.1. Capturing Synthetic Aβ Using Immobilised Metal Affinity Capture

1. ProteinChip arrays: IMAC30 (Bio-Rad; CAT # C57-30078).
2. Bioprocessors (Bio-Rad, Hercules, CA).
3. Aβ peptides.
4. Metal solution: 0.1 M solution of $CuSO_4$ (see Notes 1–2).
5. Binding buffer: Tris-buffered saline (TBS), 0.02 M Tris–HCl, pH 7.5, 0.15 M NaCl.
6. Blocking buffer: 0.1 M NaAc, pH 4.0.
7. 0.1 M hydroxyethyl piperazine ethane sulfonic acid (HEPES), pH 7.2.
8. Acetonitrile (ACN), HPLC grade.
9. Trifluoroacetic acid (TFA), HPLC grade.
10. Isopropyl alcohol (IPA), HPLC grade.
11. Energy absorbing matrix (EAM): SPA.
12. SELDI-TOF MS ProteinChip reader (Bio-Rad).
13. Plate-rotor compatible centrifuge.

2.2. Capturing Aβ from a Transgenic Mouse Brain Using W02 Antibody Capture

1. PS10 ProteinChip arrays (Bio-Rad).
2. Bioprocessors (Bio-Rad).
3. W02 antibodies: anti-human Aβ in PBS (0.25 mg/mL) (see Note 3).
4. Homogenised mouse brain.
5. Urea, 8 M.
6. Phosphate-buffered saline (PBS): 0.137 M NaCl, 0.01 M phosphate buffer, pH 7.4.
7. PBST: PBS, 0.5% Triton X-100.
8. Deactivation buffer: 0.5 M ethanolamine in PBS, pH 8.0.
9. HEPES.
10. SPA EAM.
11. TFA.
12. ACN.
13. SELDI-TOF MS ProteinChip reader (Bio-Rad).
14. Plate-rotor compatible centrifuge.

3. Methods

3.1. Capturing Synthetic Aβ Using Immobilised Metal Affinity Capture

3.1.1. Activating the IMAC30 Arrays

1. Place IMAC30 array into a bioprocessor.
2. Load each spot with 20 μL of 0.1 M $CuSO_4$ and incubate on a shaking table for 5 min. Repeat (see Note 4).
3. Load each spot with 20 μL of NaAc, pH 4, and again incubate on the shaking table for 5 min. This step aims to block non-specific binding sites on the IMAC30 array.

3.1.2. Addition of the Aβ Peptides

1. Wash spots with 20 μL of HEPES before incubating the array for 5 min on the shaking table. Repeat.
2. 20 μL of the binding buffer, TBS, is then added to each spot and the chip is again incubated on the shaking table for 5 min. The addition of the binding buffer aims to create an environment on the chip surface which is conducive to interactions between the metal and the Aβ peptides. Repeat (see Note 5).
3. 20 μL of the Aβ peptides are then added to each spot and the bioprocessor is spun in the centrifuge for 1 min, at 41 × *g* to enable the peptides to interact with the chip surface. The bioprocessor is subsequently covered with parafilm and allowed to incubate on the shaking table for 60 min (see Note 6).

3.1.3. Post-metal-Peptide Binding Washes

1. Excess sample is removed by inverting the bioprocessor before each spot is washed with 100 μL of TBS. This is followed by an incubation period on the shaking table for 1 min. Repeat (see Note 7).
2. Each spot is then washed with 100 μL of HEPES and incubated for 1 min on the shaking table. Repeat (see Note 8).
3. Subsequent to the post-metal-peptide binding washes, the array is removed from the bioprocessor and air dried. This can take from 5 to 10 min which allows adequate time for the preparation of the EAM.

3.1.4. Preparation and Addition of the Energy Absorbing Matrix

1. 200 μL of 1% TFA (aq), 170 μL ACN, and 30 μL IPA are combined to create "matrix preparation solution" (see Note 9).
2. 200 μL of the matrix prep solution is added to 5 mg of SPA to create a saturated SPA solution (see Note 10).
3. The SPA solution is vortexed for 15 s.
4. SPA solution is spun for 30 s at 3,500 × *g*.
5. Combine 100 μL of the SPA solution with 100 μL of the matrix prep solution to make 200 μL of a 50% saturated SPA solution. This step is to be performed using a dark eppendorf tube as the SPA EAM is light sensitive (see Note 11).

6. Pipette 1 μL of the 50% saturated SPA solution onto each spot and allow the array to dry. Repeat.
7. Collect data using ProteinChip reader.

3.2. Capturing Aβ from a Transgenic Mouse Brain Using W02 Antibody Capture

3.2.1. Preparation and Storage of Transgenic Mouse Brain Homogenate

PBS-insoluble transgenic mouse brain homogenate should be removed from storage, at −80°C, directly before use and should be kept on ice throughout the experiment. The brain homogenate samples are diluted 1:10 with PBST and urea, 8 M. The addition of the non-ionic detergent and denaturant, facilitate the breakdown of protein–protein and membrane–protein interactions, as well as facilitating the exposure of the antigenic epitopes on the Aβ peptides.

3.2.2. Loading the Anti-Aβ Antibody

1. Aliquot 2 μL of the anti-Aβ antibody W02 or 4G8 to each spot of a PS10 ProteinChip array (0.25 mg/mL).
2. Place array in a humidity chamber and incubate overnight, at 4°C. The extended incubation time enables the desired antibody to form covalent interactions with the array surface.
3. After incubation, excess antibody is removed and 10 μL of the deactivating buffer is loaded onto each spot. This step ensures that active sites on the array surface which are not bound to our antibody are not available to form non-specific interactions with other proteins in our sample.
4. Incubate the array in the humidity chamber for 60 min, at room temperature.

3.2.3. Loading the Biological Sample

1. Remove the array from the humidity chamber and discard the deactivation buffer. Then, place the arrays into a bioprocessor before washing each spot with 20 μL of PBST. Incubate arrays for 5 min on the shaking table. Repeat.
2. Arrays are then washed with 20 μL of PBS and again incubated for 5 min on the shaking table. Repeat.
3. 180 μL of the brain homogenate is added to each spot on the array before it is spun for 2 min at 41 ×g.
4. The bioprocessor is then covered with parafilm and placed on a shaking table to incubate for 3 h at room temperature. The extended incubation allows time for the interactions between the Aβ peptides, in the sample, and the antibody to occur. The shaking table enables the sample to remain relatively homogeneous throughout this incubation.

3.2.4. Post-antigen Binding Washing

1. Post-antigen binding, the excess sample is removed by inverting the bioprocessor over an absorbent towel before 200 μL of PBST is added to each spot. The array is again incubated on the shaking table for 5 min. Repeat (see Note 7).

2. The PBST is then washed from the array by adding 200 μL of PBS to each spot. Again the array is incubated on the shaking table for 5 min. Repeat.
3. The final wash involves the addition of 200 μL of HEPES to each spot on the array before it is once more incubated for 5 min on the shaking table. Repeat.
4. Subsequent to these washes, the excess HEPES is removed and the array is removed from the bioprocessor and allowed to dry in preparation for the EAM addition. This should take between 5 and 10 min at room temperature (see Note 8).

3.2.5. Preparation of the SPA EAM

For brain samples, the preparation of the SPA EAM is consistent with that described in Subheading 3.1.4 with the exception of step 1, this should instead read:

1. 200 μL of 1% TFA (aq) and 200 μL of ACN are combined to create "matrix prep solution".

3.2.6. Addition of the SPA EAM

1. Pipette 1 μL of the 50% saturated SPA solution onto each spot and allow the array to dry. Repeat.
2. Collect data using ProteinChip reader.

3.3. Parameters for the Acquisition of Spectral Data

3.3.1. Calibrating the ProteinChip Reader

Prior to the acquisition of the Aβ mass spectra, it is imperative to calibrate the ProteinChip reader to ensure that the intensities and masses acquired are accurate. Calibration should be performed using a set of standard calibrants which cover the mass range of the peptide of interest, in the case of Aβ the mass range of interest is between 4 and 5 kDa for the monomeric species. However, as oligomeric species may be observed, it is often important to have a wider range of calibrants. A desired standard peptide calibration includes seven peptides ranging from 1 kDa (Arg8-vasopressin) to 12 kDa (Cytochrome C) and provides a robust calibration range for the acquisition of monomeric Aβ as well as small oligomeric Aβ species.

3.3.2. Acquisition of Aβ Spectral information

The calibration of the ProteinChip reader provides an approximation of the laser intensity settings required to acquire accurate Aβ spectra (see Note 12) However, it is still important to optimise the laser settings on a small subset of the array for each separate experimental procedure. This ensures that peaks within the spectra maintain a high level of resolution and reproducibility throughout the experiment (see Note 13).

Examples of spectra obtained from antibody capture using biological samples and IMAC using synthetic $A\beta_{1\text{-}40}$ are shown in Fig. 1a, b, respectively.

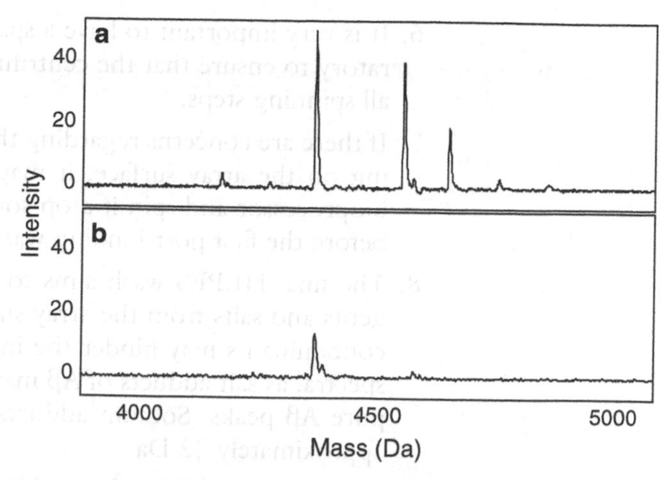

Fig. 1. Representative surface-enhanced laser desorption ionisation time-of-flight mass spectrometry spectra of monomeric Aβ species. (**a**) Captured from biological samples using W02 antibody capture. (**b**) Synthetic $A\beta_{1-40}$ captured using immobilised metal affinity capture.

4. Notes

1. While copper was used in the current method, other metal solutions, such as $FeCl_3$ or $NiCl_2$, may be used in place of $CuSO_4$ to produce similar results. These metals may be more effective when attempting to capture non-amyloid peptides or when using IMAC with biological samples.
2. As the metals utilised in IMAC investigations are transition metals, it is important to make fresh metal solutions before each experiment. This is to avoid oxidation of the solutions and the subsequent formation of hydroxyl radicals.
3. It is important to guarantee that the antibodies are stored in a manner which is free from sodium azides and any solvents containing free amines. This is to avoid competition between these substances and the antibodies for the active sites on the PS10 array surface.
4. It is important to ensure that the metal solutions are in contact with the array surface. This may be ensured by spinning the bioprocessor in the centrifuge for approximately 30 s at 41 ×*g*.
5. PBS may also be used as the binding buffer in these steps. However, when using metals, such as iron, which are known to interact with phosphorylated peptides, it is best to avoid using buffers that contain phosphate, and thus TBS is often utilised for these wash steps.

6. It is very important to have a spare bioprocessor in the laboratory to ensure that the centrifuge remains balanced during all spinning steps.
7. If there are concerns regarding the volume of sample remaining on the array surface, it may be beneficial to invert the bioprocessor and spin it atop some absorbent towel for 30 s before the first post-binding wash.
8. The final HEPES wash aims to remove any traces of detergents and salts from the array surface. The presence of these contaminants may hinder the interpretation of the resulting spectra, as salt adducts of Aβ may be observed in the place of pure Aβ peaks. Sodium adducts of Aβ have a mass shift of approximately 22 Da.
9. The addition of IPA allows the matrix to dry at a faster rate and thus affects matrix crystallisation. Consider altering the volume of IPA, or any other organic solvent, in an effort to optimise matrix crystallisation for each experiment.
10. SPA EAM may be replaced with α-cyano-4-hydroxy cinnamic acid (CHCA) or 2,5-dihydroxybenzoic acid. These matrices crystallise in different manners and thus result in the production of different spectral qualities. It may be beneficial to test different matrices in an effort to optimise the resulting spectra.
11. When adding 100 μL of the SPA solution, be careful to avoid adding SPA precipitate from the bottom of the eppendorf tube. If required respin the solution. Addition of solid SPA to the matrix mix may affect matrix crystallisation.
12. The pre-made standard peptide arrays, available from Bio-Rad, are loaded with CHCA EAM. Hence, when using SPA EAM during experiments a higher laser setting might be required for ionisation to achieve peak intensities equivalent to those of the peptide standards. This is due to the difference in energy absorption between CHCA and SPA.
13. As mentioned by Giannakis et al. (41), it may be beneficial to have distinct data acquisition settings for low molecular mass (LMM) species and high molecular mass (HMM) species, respectively. Peptide species of differing masses require different acceleration energies and thus the settings used in LMM data acquisition may not be appropriate for collecting data from within the HMM range.

References

1. Gravina, S.A., et al., *Amyloid β protein (A β) in Alzheimer's disease brain. Biochemical and immunocytochemical analysis with antibodies specific for forms ending at Aβ40 or Aβ42(43).* Journal of Biological Chemistry, 1995. **270**: p. 7013–7016.
2. Kosik, K.S., C.L. Joachim, and D.J. Selkoe, *Microtubule-associated protein tau (τ) is a major*

antigenic component of paired helical filaments in Alzheimer disease. Proc Natl Acad Sci U S A, 1986. **83**: p. 4044–4048.

3. Chetelat, G., et al., *Direct voxel-based comparison between grey matter hypometabolism and atrophy in Alzheimer's disease.* Brain, 2008. **131**(1): p.60–71.
4. Masters, C.L., et al., *Amyloid plaque core protein in Alzheimer disease and Down syndrome.* Proc Natl Acad Sci U S A, 1985. **82**: p. 4245–4249.
5. Glenner, G.G. and C.W. Wong, *Alzheimer's disease: initial report of the purification and characterization of a novel cerebrovascular amyloid protein.* Biochem Biophys Res Commun, 1984. **120**: p. 885–890.
6. Goldgaber, D., et al., *Isolation, characterization and chromosomal localization of human brain cDNA clones coding for the precursor of the amyloid of brain in Alzheimer's disease, Down's syndrome and aging.* Journal of Neural Transmission, 1987. **24**: p. 23–28.
7. Kang, J., et al., *The precursor of Alzheimer's disease amyloid A4 protein resembles a cell-surface receptor.* Nature, 1987. **325**: p. 733–736.
8. Robakis, N.K., et al., *Molecular cloning and characterization of a cDNA encoding the cerebrovascular and neuritic plaque amyloid peptides.* Proc Natl Acad Sci U S A, 1987. **84**: p. 4190–4194.
9. Tanzi, R.E., et al., *Amyloid β protein gene: cDNA, mRNA distribution and genetic linkage near the Alzheimer locus.* Science, 1987. **235**: p. 880–884.
10. Blennow, K., M.J. de Leon, and H. Zetterberg, *Alzheimer's disease.* Lancet, 2006. **368**(9533): p. 387–403.
11. Zhang, Y.-W. and H. Xu, *Molecular and Cellular Mechanisms for Alzheimer's Disease: Understanding APP Metabolism.* Current Molecular Medicine, 2007. 7(7): p. 687–696.
12. Prelli, F., et al., *Different processing of Alzheimer's β-protein precursor in the vessel wall of patients with hereditary cerebral hemorrhage with amyloidosis - Dutch type.* Biochemical and Biophysical Research Communications, 1988. **151**: p. 1150–1155.
13. Bica, L., et al., *Metallo-complex activation of neuroprotective signalling pathways as a therapeutic treatment for Alzheimer's disease.* Molecular Biosystems, 2009. **5**(2): p. 134–142.
14. Masters, C.L., et al., *Molecular mechanisms for Alzheimer's disease: implications for neuroimaging and therapeutics.* J Neurochem, 2006. **97**(6): p. 1700–1725.
15. Jellinger, K.A., et al., *Biomarkers for Early Diagnosis of Alzheimer disease: 'ALZheimer ASsociated gene' - A New Blood Biomarker?* Journal of Cellular and Molecular Medicine, 2008. **12**(4): p. 1094–1117.
16. Crouch, P.J., A.R. White, and A.I. Bush, *The modulation of metal bio-availability as a therapeutic strategy for the treatment of Alzheimer's disease.* FEBS Journal, 2007. **274**(15): p. 3775–3783.
17. Bush, A.I. and R.E. Tanzi, *Therapeutics for Alzheimer's disease based on the metal hypothesis.* Neurotherapeutics, 2008. **5**(3): p. 421–432.
18. Massie, H.R., V.R. Aiello, and A.A. Iodice, *Changes with age in copper and superoxide-dismutase levels in brains of C57BL-6J mice.* Mechanisms of Ageing and Development, 1979. **10**(1–2): p. 93–99.
19. Maynard, C.J., et al., *Overexpression of Alzheimer's disease amyloid-β opposes the age-dependent elevations of brain copper and iron.* J Biol Chem, 2002. **277**(47): p. 44670–6.
20. Takahashi, S., et al., *Age-related changes in the concentrations of major and trade elements in the brain of rats and mice.* Biological Trace Element Research, 2001. **80**(2): p. 145–158.
21. Barnham, K.J., C.L. Masters, and A.I. Bush, *Neurodegenerative diseases and oxidative stress.* Nature Reviews Drug Discovery, 2004. **3**(3): p. 205–214.
22. Barnham, K.J., et al., *Metal-Protein Attenuating Compounds (MPACs) for the Treatment of Alzheimer's Disease.* Drug Design Reviews-Online, 2004. **1**: p. 75–82.
23. Finefrock, A.E., Bush, A.I., and P.M. Doraiswamy, *Current status of metals as therapeutic targets in Alzheimer's disease.* J Am Geriatr Soc, 2003. **51**(8): p. 1143–1148.
24. Atwood, C.S., et al., *Characterization of copper interactions with Alzheimer amyloid β peptides: identification of an attomolar-affinity copper binding site on amyloid β1-42.* J Neurochem, 2000. **75**: p. 1219–1233.
25. Opazo, C., et al., *Metalloenzyme-like activity of Alzheimer's disease β-amyloid: Cu-dependent catalytic conversion of dopamine, cholesterol, and biological reducing agents to neurotoxic H_2O_2.* J Biol Chem, 2002. **277**(43): p. 40302–40308.
26. Adlard, P.A., et al., *Rapid restoration of cognition in Alzheimer's transgenic mice with 8-hydroxy quinoline analogs is associated with decreased interstitial Abeta.* Neuron, 2008. **59**(1): p. 43–55.
27. Crouch, P.J., et al., *Increasing Cu bioavailability inhibits A beta oligomers and tau phosphorylation.* Proceedings of the National Academy of Sciences of the United States of America, 2009. **106**(2): p. 381–386.
28. Hanger, D.P., et al., *Glycogen synthase kinase-3 induces Alzheimer's disease-like phosphorylation of tau: Generation of paired helical filament*

epitopes and neuronal localisation of the kinase. Neurosci Lett, 1992. **147**(1): p. 58–62.

29. White, A.R., et al., *Degradation of the Alzheimer disease amyloid beta-peptide by metal-dependent up-regulation of metalloprotease activity.* Journal of Biological Chemistry, 2006. **281**(26): p. 17670–17680.
30. Issaq, H.J., et al., *SELDI-TOF MS for diagnostic proteomics.* Analytical Chemistry, 2003. **75**(7): p. 148A–155A.
31. Whelan, L.C., et al., *Applications of SELDI-MS technology in oncology.* Journal of Cellular and Molecular Medicine, 2008. **12**(5A): p. 1535–1547.
32. Hutchens, T.W. and T.T. Yip, *New Desorption Strategies for the Mass-Spectrometric Analysis of Macromolecules.* Rapid Communications in Mass Spectrometry, 1993. 7(7): p. 576–580.
33. Merchant, M. and S.R. Weinberger, *Recent advancements in surface-enhanced laser desorption/ionization-time of flight-mass spectrometry.* Electrophoresis, 2000. **21**(6): p. 1164–1177.
34. Wittke, S., T. Kaiser, and H. Mischak, *Differential polypeptide display: the search for the elusive target.* Journal of Chromatography B-Analytical Technologies in the Biomedical and Life Sciences, 2004. **803**(1): p. 17–26.
35. Vorderwülbecke, S., et al., *Protein quantification by the SELDI-TOF-MS-based Protein-Chip® System.* Nature Methods, 2005. **2**(5): p. 393–395.
36. Geng, T., et al., *Use of surface enhanced laser desorption/ionization-time of flight mass spectrometry (SELDI-TOF MS) to study protein expression in a rat model of cocaine withdrawal.* Journal of Neuroscience Methods, 2006. **158**(1): p. 1–12.
37. Bio-Rad Laboratories, I., *ProteinChip IMAC30 Array (Immobilized Metal Affinity Capture) - Instruction Manual*, I. Bio-Rad Laboratories, Editor. 2006.
38. Qin, W., et al., *S100A7, a Novel Alzheimer's Disease Biomarker with Non-Amyloidogenic α-Secretase Activity Acts via Selective Promotion of ADAM-10.* PLoS ONE, 2009. **4**(1): p. e4183.
39. Simonsen, A.H., et al., *Identification of a novel panel of cerebrospinal fluid biomarkers for Alzheimer's disease.* Neurobiology of Aging, 2008. **29**(7): p. 961–968.
40. Hung, L.W., et al., *Amyloid-beta peptide (Abeta) neurotoxicity is modulated by the rate of peptide aggregation: Abeta dimers and trimers correlate with neurotoxicity.* J Neurosci, 2008. **28**(46): p. 11950– 11958.
41. Giannakis, E., et al., *Analysis of Aβ Interactions Using ProteinChip Technology.* Methods Mol Biol, 2008. p. 71–86.

INDEX

Andrew F. Hill et al. (eds.), *Protein Folding, Misfolding, and Disease: Methods and Protocols*, Methods in Molecular Biology, vol. 752, DOI 10.1007/978-1-60327-223-0, © Springer Science+Business Media, LLC 2011

B

C

I

K

L

M

N

O

P

Printed in the United States
By Bookmasters